CAMBRIDGE COUNTY GEOGRAPHIES

SCOTLAND

General Editor: W. Murison, M.A.

DUMFRIESSHIRE

Cambridge County Geographies

DUMFRIESSHIRE

by

JAMES KING HEWISON, M.A., D.D.

Fellow of the Society of Antiquaries of Scotland

With Maps, Diagrams and Illustrations

Cambridge :

at the University Press

1912

CAMBRIDGE UNIVERSITY PRESS
Cambridge, New York, Melbourne, Madrid, Cape Town,
Singapore, São Paulo, Delhi, Mexico City

Cambridge University Press
The Edinburgh Building, Cambridge CB2 8RU, UK

Published in the United States of America by Cambridge University Press, New York

www.cambridge.org
Information on this title: www.cambridge.org/9781107623408

First published 1912
First paperback edition 2013

A catalogue record for this publication is available from the British Library

ISBN 978-1-107-62340-8 Paperback

CONTENTS

ILLUSTRATIONS

ILLUSTRATIONS

The illustrations on pp. 5, 9, 13, 17, 22, 25, 51, 52, 64, 68, 74, 108, 117, 118, 122, 123, 127, 132, 134, 151, 154, 162, 165, 168, 170, and 171 are from photographs by Messrs J. Valentine & Sons; those on pp. 19, 35, 37, 40, 41, 94, 102, 104, 105, 110, 113, 116, 120, 121, 156, 158, 160, and 166 are from photographs by the author; the portraits on pp. 137, 142, and 148 are from photographs by Messrs T. & R. Annan; the portrait on p. 139 is from an etching by kind permission of Sir George Reid; that on p. 146 is from a photograph by Mr John Fergus; the illustration on p. 159 is from a photograph by Miss Montgomerie, Dalmore; the illustrations on pp. 87 and 98 are reproduced by courtesy of the Society of Antiquaries of Scotland; that on p. 144 is from a block kindly lent by Messrs J. Maxwell & Son, Dumfries; those on pp. 23 and 45 are reproduced from Mr Hugh S. Gladstone's *Birds of Dumfriesshire* by kind permission of the author, the former being from a photograph by Mr Legard; that on p. 31 is reproduced from Geikie's *Scenery of Scotland* by permission of Messrs Macmillan & Co. The Map of the Solway on p. 48 is reproduced by courtesy of Dr George Neilson.

1. County and Shire. The Origin of Dumfries.

The creation of a county and the establishment of a sheriffdom in Dumfriesshire were fraught with difficulties. The ancient county of Dumfries included part of Galloway as far west as the river Cree, in addition to the present area which was constituted a sheriffdom in 1748. When in 1107 King Edgar bequeathed to his youngest brother, Prince David, Scottish Cumbria, of which the present shire of Dumfries was then a part, he granted a very disputable possession. It was a little buffer-state between two warring kingdoms. David, being both a petty king and earl (*comes*), had the opportunity for imposing upon his territory the feudal system, of the Anglo-Norman type, to which he had been accustomed in England. To his court he attracted Anglo-Norman and southern chivalry to support him in his rule. His regal administration was probably conducted by feudal dignitaries—chancellor, constable, justiciar, chamberlain, steward, and marshal. On David's accession to the throne of Scotland in 1124, according to Gaelic custom and feudal law, his personal property became an appanage of the crown.

But at least one of the three great divisions of the border-land, namely Strathnith, was still ruled by a Celtic over-lord Dunegal; and, in like manner, probably Annandale and Eskdale were governed by hereditary chiefs. David found it impolitic at once to discard the old code of law and customs, which, as in the case of Galloway, prevailed in some measure for centuries. The eastern boundary of Gaway or Galloway is not easily determined now. Consequently the Celtic over-lord of Strathnith (Nithsdale, and probably part of Galloway) was left undisturbed. Annandale, however—a tract stretching to the Forest of Selkirk—was granted to Robert de Brus, while the constable, Morville, got Cunningham, and the steward, FitzAlan, got Renfrew and part of Kyle. To prevent jealousies among the local chiefs the Brus was not created an earl. If Dunegal held the office of a *Maor* (who corresponded to the *Gerefa* or sheriff of the Saxons) on his own land, he might act as a *vice-comes*. The early kings themselves, in their progresses with their justiciars, presided over the courts of law. It is natural to expect, therefore, that the castle-guard of the county was connected with the territory governed by Dunegal. The county (*comitatus*) for seven centuries has been associated with the town of Dumfries—a place where Dunegal and Radnulf his son held and disponed heritage about the middle of the twelfth century. Radnulf's charter was given at "Dronfres," which in the Gaelic tongue signifies "the ridge of the bushes" (*phreas*). This corresponds with the persistent local pronunciation "Drumfreesh." The next form of the word is *Dunfres* and *Dunfrez* (1183–8), a significant change after the *dun*, or fort of Dunegal, on the bushy ridge, became of paramount importance. This form of the word, "Dunfrys," appears in 1296, and "Drumfres" holds on in charters after 1329. What in the way of establishing feudalism David and Malcolm left undone William the Lyon completed. About the year 1186 he erected a strong castle (*castrum*) at Dunfres to overawe the rebellious Galwegians. In that military centre the officers of the king held their courts, and received the service and fees of the knights and barons of the district as well as customs due to the Crown. Annandale was exempted from castle-ward. Guarded by the castle, the new royal burgh of Dumfries thus early rose, and from it the sheriff made the lieges keep the king's peace. The boundaries of this sheriffdom (*vice-comitatus*) in course of time were curtailed. The sheriff's jurisdiction, however, was not, in all matters, commensurate with the boundaries of the county. Annandale had the separate jurisdiction of a stewartry, Eskdale that of a regality. Till the middle of the eighteenth century, sheriffdom, stewartry, and regality were hereditary in different noble families in succession. When the Scotts of Buccleuch held the regality, they got their Eskdale lands transferred to the sheriffdom of Roxburgh, from which they were separated in 1748 and joined again to Dumfries. The burgh of Dumfries was in the fifteenth century excluded from the sheriff's jurisdiction in respect of "actions of blood"—a privilege confirmed to the town by James IV in 1509.

The county became adjusted to its modern conditions

when parliament, by the "Heretable Jurisdictions" Act, 1747, abolished heritable jurisdictions. The Duke of Queensberry, hereditary sheriff of Dumfries and coroner of Nithsdale, received £6621. 8s. 5d. in compensation for his loss of offices; the Marquis of Annandale, £3000 for the loss of the stewardship of Annandale, and the regality of Moffat; and the Duke of Buccleuch £1400 for the regality of Eskdale.

2. General Characteristics.

In the southern uplands of Scotland, stretching between sea and sea, where the two kingdoms meet in the Solway Strath, and almost encompassed with an oval girdle of green hills, lies the pastoral territory of Dumfries. The picturesque aspect of the country made Fergusson, the poet, declare that the gods "there ha'e shown their power in fairy dream." Its characteristics are varied, no one having a striking predominance, since the verdant landscape, from its maritime margin on the Solway Firth, stretches evenly over holm and undulating ground, ridges and little hills, up to a high transverse watershed, ranging in height from Corsincone Hill (1547 feet) to White Coomb (2695 feet) above sea-level. Nothing appears exaggerated, and a fascinating harmony everywhere prevails, partly created by landscape artists who have blended woodland and tilled field to add beauty to hill, dale, and river.

Three great vales, parallel to each other, and having a southerly trend—Nithsdale, Annandale and Eskdale—each drained by a river from which it is named, give the shire its chief natural distinction. In combination with these are lesser dales—the greatest being the Vale of Cairn. These again slope down on either side of the three great water arteries, in two directions, south-west and south-east respectively; and thus nature provides for moderating the wind, rain, and fog, and for obtaining an even distribution

The Castle Loch, Lochmaben

of sunshine, heat, and moisture. The valleys, watered by a hundred trouting-streams, which drain much arable and pastoral land, form suitable tracts for the agriculturist and sheep-farmer. In early autumn, when the harvest crowns the year, the prospect from one of the great hills is a vast panorama of green and gold cut with a streak of silver river in each of the valleys below.

On the uplands there are wide tracts of moorland and hill pasture, lone, yet vocal with the bleat of sheep and cry of wild birds. Lakes are little in evidence except around Lochmaben—"Marjory o' the mony lochs, A carlin auld and teuch," where formerly seven sheets of water were clustered together. Worthy of note is an extensive tract of peat moss and moor lying east of and parallel to the river Nith, below the town of Dumfries. Along its eastern margin the G. & S. W. Railway is built. Known as Lochar Moss, it has the divisional names of Craig's Moss, Racks Moss, Lochar Moss, Ironhirst Moss, Holmhead Moss, and Longbridge Moor. In extent it is six miles long and above two miles broad at the Wath Burn. Recently extensive plant has been built on the moss for the utilisation of the peat.

The maritime position of the shire is also of considerable importance still, notwithstanding the greater convenience of the railways for the carriage of imports and exports. Nith and Annan are deep tidal rivers with water to bring vessels of considerable tonnage to near Dumfries and to Annan. Nature has provided an easy outlet, by the rivers Nith and Annan and by the Solway Firth, for the exportation of the local minerals, products, manufactures, and netted fish. Unfortunately it is still the upper surface of the land which is the source of wealth and power. The soil being utilised to the fullest extent affords work and homes for a busy population which now has little chance to increase much in country areas.

One noticeable feature of the landscape for which man is wholly responsible is the regular division of arable and grass lands into fields, well fenced with stone dykes, with hedges of intermingled thorn and beech, and with iron fences. Shapely and well-assorted plantations of soft and hard wood indicate a careful land culture. Great herds and flocks move on numerous pastures. Everywhere comfortable farmsteads appear. On prominent spots many hoary ruined fortalices stand, and towers, repaired and enlarged into modern mansions, are visible in wooded demesnes in greater numbers than in other shires. These residences indicate the presence of a class personally interested in the land, being heirs or successors of the many barons and proprietary who once were powerful factors in the affairs of The Borders.

3. Size. Shape. Boundaries.

In size Dumfriesshire ranks eighth among Scottish counties, and in population fourteenth. Its greatest length from north-west to south-east, that is, from Ellergoffe Knowe (1264) to Liddel, is 53½ miles, and from north to south-south-west, that is, from Loch Craig Head (2625) to a point south-south-west of Caerlaverock Castle, 32⅜ miles. The shore line extends 32 miles. Its area contains 708,071 acres, inclusive of foreshore and tidal water, or 1106 square miles ; or 690,294 acres exclusive of these, or 1078 miles. There are 4000 acres, or six square miles, of inland water.

This area would appear on the map as an oval with a serrated boundary, were the northern indentation by

Lanarkshire removed. Six lowland Scottish shires—
Kirkcudbright, Ayr, Lanark, Peebles, Selkirk and Roxburgh, and one English county—Cumberland—gird the
shire.

Queensberry Hill (2285) stands near the centre of the
northern boundary. From it to either side springs a great
arc of high hills, thereby forming two almost semicircular
boundaries along the northern frontier. Beginning the
circuit on the extreme west side, where Dumfries touches
Kirkcudbright and Ayr, one finds the Lorg range, with
three summits, rising in Blacklorg to 2231 feet, and forming a watershed for Afton on the north and the many
feeders of the Nith in the west. A lofty ridge curves
around Kirkconnel by McCririck's Cairn (1824) beneath
Corsincone (1547), on to the limits of Ayr and Lanark
at Threeshire Stone (1500), near Mount Stuart Hill
(1567), then eastward by Spango (1391) to Wanlock Dod
(1808), between Wanlockhead and Leadhills. Thence
the watershed sweeps south to Lowther Hill (2377)—the
highest altitude there. The range still keeps high while
bounding Durisdeer, in the Waal Hill reaching 1987 feet,
and in Wedderlaw 2185 feet.

Beginning again slightly north of Queensberry, the
course of the second arc tends northward with a declivity
of a few hundred feet till it touches Peeblesshire and
Lanarkshire, beneath Flecket Hill (1522), in the vicinity
of Annanhead Hill (1566), and the Devil's Beef Tub.
Again ascending to the north-west with increasing altitude
—at Hartfell (2651)—it comes to the boundary of Selkirk-
shire at Loch Craig Head (2625), 70 feet lower than

The Devil's Beef Tub

White Coomb (2695).   The boundary curves round Loch
Skene; and taking a southerly course at a height of over
2000 feet, it reaches Capel Fell (2223), east of Moffat
Water.   A northern sweep by Ettrick Pen (2270)—the
highest hill in that quarter—brings it with varying declivity
round Eskdalemuir, south of Ettrick, past the boundary
of Roxburghshire, fixed above Moodlaw Loch, to Cause-
way Grain Head (1607).   It then descends southward by
Hartgarth Fell (1806), east of Ewes, till it meets Liddel
Water, the boundary with England.

Liddel Water, till it joins the Esk near The Scots
Dyke, for five miles in straight line from point to point,
forms the boundary between Canonbie and Cumberland.
The Scots Dyke runs till it almost touches the river Sark,
which now becomes the boundary between the countries
as it flows south to join the Esk.   In the channel of the
Esk, as it flows west, the boundary is fixed for a few miles,
till Esk reaches the Eden opposite Torduff Point.   Eden
falls into Solway; and Solway Firth, below Newbie,
becomes the extreme boundary on the south.

To complete the environment—going north from this
point the boundary lies between Dumfries and Kirkcudbright till Ayrshire is reached at Meikledodd Hill (2100).
At first it divides Blackshaw Bank and does not touch the
Nith till well up its channel, which it crosses and rejoins
beneath Kelton Bank.   In the Nith, flowing between
Dumfries and Maxwelltown, the boundary is fixed till it
reaches the Water of Cluden at Lincluden, whence it goes
in a north-westerly direction till Lower Stepford is reached.
Here the Barbuie Burn separates the shires, and the

boundary runs south of Dunscore parish between hills rising over 1000 feet, on past Craigenputtock (700) and Loch Urr, round Glencairn, across Benbrack (1900) and Black Hill (1808)—the western limit of Tynron parish. Touching Penpont, the boundary next reaches to Sanquhar at Blacklorg (2231)—the western limit as stated at the outset.

In the thirteenth century and onwards the boundary between England and Scotland was the river Esk. In 1552 the frontier was shifted to the Sark. The district between the old and new limits was debateable. This Debateable Land, known as "the lands Batable or Threpe Lands," lay partly in England and partly in Scotland. Its south boundary was the Esk from its junction with the Liddel to the spot where the Esk and the Sark met. It comprehended the baronies of Kirkandrews and Morton in Cumberland, and Brettalach or Bryntalone, now Canonbie, in Dumfriesshire.

4. Surface and General Features.

The surface of the county, rising from sea-level to an altitude of nearly 2700 feet, is varied with holm and ridge, dell and hill, with here and there a sheet of water breaking the monotony of a green landscape. Its pastoral aspect is noteworthy. One-half of the whole area is covered with hill-grass and heath, more than one-fifth with permanent pasture, and more than one-fifth is arable land. Thus there is little left for woodland, water, and waste in a territory agreeably parcelled out into 2700 agricultural holdings.

The surface, generally speaking, presents no abrupt features, but meadow land slopes up to ridge, ridges reach to uplands, and these rise to the high spurs which emerge from the barriers on the northern limits. The systematic disposition of the great valleys and minor dales, already referred to, with their relative streams, forms a certain regularity of surface in every district. This lie of the land permits nature to display every charm from the beauty and riches of long and well-tilled alluvial fields of a red hue, the picturesque aspect of copse and wood, self-sown or planted by river and bank up to the hills clad with heath or mountain grass. The upper valley of Nith, the middle basin round Thornhill, and the ever widening Strath extending from Dumfries to Solway Firth are in turn scenes of pastoral and sylvan beauty. Similarly the triple vales of Annan, narrow at Moffat, widening into "The Howe of Annandale," and debouching into the Solway Strath, have characteristics almost identical with those of Nithsdale. Eskdale, with its three vales of Ewes, Esk, and Wauchope, united at the town of Langholm, less extensive and more rural, still beautiful, sooner descends into the Lowlands.

Of the lofty hills piled in concentric circles along the northern frontiers enough has been written to suggest the grandeur of many serene vales at their base, and many noisy torrents in rocky beds, such as Crichope Linn, and the Cauldron at Wamphray in the hills. Travellers have felt the charm of the pensive Passes of Mennock, Dalveen, and Waalpath, and the glamour of Scaur Water and Shinnel in Nithsdale: have realised the grandeur investing The Devil's Beef Tub in Annandale, and lonely Loch Skene in the wilds of Moffatdale: and have lingered on Langholm Bridge, Whita Hill, Hollows Tower, and other fascinating spots in Eskdale.

Of prominent hills in the west is Cairnkinna (1819)

Dalveen Pass

in Penpont parish, overlooking Nithsdale, Glencairn, and the Solway Strath as far as Cumberland. Queensberry (2285), in the very centre, commands Nithsdale, Annandale, the seaboard and the silvery Firth beyond. Between these points, eminent stands Drumlanrig Castle (300) surmounting the landscape garden of Middle Nithsdale made by Douglases and Scotts.

From Hartfell (2651) in Moffat parish, nearly 366

feet higher than Queensberry, the Firth and the German Ocean can be seen.

Eskdale and the Border land for many miles is dominated by Whita Hill (1162), on which is erected an obelisk in memory of Sir John Malcolm. The Roman strategists proved their knowledge of surveying when they established Birrenswark Camp (920), in Hoddom parish, as a frontier post from which beacon signals could be communicated far on every side.

Perhaps the beautiful holm lands in all the vales, when in pasture or crop, impress a visitor most; but the happy distribution of waters running in romantic courses shaded with luxuriant trees is a feature of the shire oft recalled. The rounded bosses of the hills too are impressive features. Extensive mosses also suggest the age when the primal forests stood in majesty, as now the last of the ancient oak woods stands in decadence, great but dying, round Lochwood ruined Tower, beside a vast peat moss.

5. Watershed. Rivers and Lakes.

The lakes are mere fountain-heads of inconsiderable streams. The true watershed is the chain of lofty hills running from west to east on the northern frontier. From this elevated plateau the water is shed into three out of the four cardinal directions of the compass. At two points, only 11 miles apart, rise the important rivers Tweed and Clyde; the former emerging west of Annanhead Hill (1566) from Flecket Hill (1522), from which also, and not half a mile away, springs a leader which feeds one of the three head-waters of the Annan.

The Clyde, in its principal head-stream—Daer Water —rises on the north side of Gana Hill (2190) on the boundary of Closeburn, four miles and a half due east of Carronbridge railway station. Both Tweed and Clyde flow north for a considerable distance then descend respectively east and west. It is different with the river Nith (55 miles), which has a south-easterly course across the watershed made for it in a local depression. Three times in its sluggish journey through the parishes of New Cumnock, Kirkconnel, and Sanquhar would slight barriers force its waters back to join the Glasnock and to emerge at Ayr. Nith, starting from Benbrack Hill (1621) in Dalmellington, flows east into New Cumnock, and runs first in a north-easterly direction, then eastward past New Cumnock town to the boundary of the shire in Kirkconnel. Increased by the Afton and other tributaries which descend from the watershed, Nith without the Ayrshire waters would still have been a river of volume like Annan. It runs into a narrow pass, its banks here and there fringed with trees, before it reaches the village of Kirkconnel, where a more open oval tract of undulating ground appears. This vale of Upper Nithsdale, with its distinct circuit of hills over 1100 feet high, is nine miles long and over four miles broad. In its descent the river on its right bank is joined below Kirkconnel by the Kello (10), and below the burgh of Sanquhar by the Euchan (9); while on the left bank above Sanquhar the romantic Crawick (8) falls into it. Passing through the woodlands of Eliock, on the left swollen by the Mennock (6), Nith finds restraint within a grey picturesque defile through which it winds past the hamlet below historic Enterkin. It next plunges into the red ravines above Drumlanrig, through its woods, passing half a mile east of the Castle, beneath which it is joined by Marr Burn. A little lower, near Carronbridge, the Carron (9), draining the dale of Durisdeer, broadens the river speeding on in serpentine track through a vast garden. Middle Nithsdale, or The Thornhill Basin, as geologists term it, is an emerald oval, eleven miles long and seven miles broad, bisected by the river Nith.

At a distance of one mile and a half south-west of Thornhill, the Cample (10), increased by streamlets from the heathery hills on the east, falls into Nith. A mile farther south, the brown waters of the Scaur (20)— increased near Penpont by its tributary the Shinnel (12)— having drained Penpont, Tynron, and part of Keir, roll over the bluish-grey gravel at Keir into Nith. Flowing on down the valley, straitened at Blackwood, through the red bridge of Auldgirth, into the Dumfries basin circled by the hills of Kirkmahoe, Dunscore, Terregles, Mabie and Tinwald, Nith, deeper and broader, continues its oblique course through this arable tract. It obtains additional volume at Lincluden Abbey, where the waters of Cluden (23), gathered by many streamlets in Dunscore and Glencairn parishes—namely Old Cluden, Glenesslin, Cairn, Castlefern, Craigdarroch, and Dalwhat, all from the west—fall into Nith. Flowing through Dumfries and Maxwelltown in a stream 94 yards broad, it meets

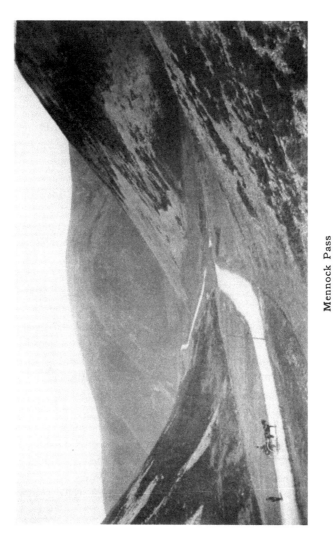

Mennock Pass

the tide and with it rushes to the Solway Firth, passing on the Dumfries side Kingholm, Kelton, and Glencaple, the local ports.

Running nearly parallel with the Nith from a source in Kirkmahoe, east of High Auldgirth, Park Burn (16)—then Lochar Water—flows south through the parishes of Kirkmahoe, Tinwald, and Torthorwald into Lochar Moss, touching Dumfries and Mouswald, where it is joined by the Wath Burn, then on through Caerlaverock into Ruthwell, where its sluggish stream falls into the Solway.

Two miles below Moffat three mountain streams unite in the holms and form the river Annan (40). North of that point at Annanhead Hill (1566) the middle stream, least of the three, the Annan is supposed to rise; but a longer feeder—the Lochan (7)—descends from Barry Grain Rig.

The second head-stream, Evan, on the west descends from the highlands of Lanarkshire through romantic, wooded defiles traversed from Beattock by the highway to Edinburgh and by the Caledonian Railway. Of its many tributaries Garpol Water is the greatest. The third and longest branch, Moffat Water (13), supposed to rise at Birkhill (1080), really rises at the extreme boundary at Loch Craig Head (2625) whose southern slopes drain into Loch Skene. Its sustaining rivulets descend through 19 "cleuchs," "grains" or "gills," stretching up to the boundaries on both sides. The coach road to St Mary's Loch is parallel to the stream.

South of the meeting of these three streams, where

Moffat : the Well Burn

the vale is three miles and a half broad, the river flows south between Wamphray and Johnstone into the widening valley of Mid Annandale, which surrounds Lochmaben and Lockerbie.   Here the oval basin contains a smaller cup—seven miles across from Quhytewoollen Hass (733) to Hightown Hill (818)—in which lie the Lochs of Lochmaben.   Before reaching this depression the river has been joined, first by the Wamphray (8) from Croft Head (2085), and below Applegarth by the united streams of Kinnel and Æ.   Kinnel (17) takes its rise in the uplands of Kirkpatrick Juxta beyond Queensberry, and, having passed through the woods of Johnstone, it descends to meet the Æ, rising on the other side of Queensberry and draining Kirkmichael parish.   South of a line drawn between Lockerbie and Lochmaben, the Dryfe (13), from the east, running from Loch Fell (2256) and draining the narrow dale of Dryfe, falls into Annan.   Now bending south-west Annan, joined by the Milk (17) from the east, in its united stream of Milk and Corrie, reaches the lovely woods of Hoddom.   Another tributary from the east is the Mein (7), from Tundergarth, which mingles with Annan at a point one mile and a half south-south-west of Ecclefechan.   In an expansive Strath, Annan passes Brydekirk village, at Annan meets the tidal waters, and two miles lower down reaches the Waterfoot.

With devious, almost semicircular course, the Kirtle (15), rising from the hills between Middlebie and Langholm, after draining Middlebie and Kirkpatrick Fleming, falls into the Sark south of Old Gretna, and near the confluence of the river Esk.

The third great valley in the county is watered by the Esk (40). The White Esk (6) rises to the north of Ettrick Pen (2270) in Eskdalemuir. The Garvald (5) from the western Fells joins the White Esk at the 700 feet level; and the united streams swelled by the Moodlaw, Rae, and other streamlets from the east, flow southward beyond Westerkirk till, on the west, they are joined by the Black Esk. The Black Esk (12) takes its rise at Jock's Shoulder (1754). Now styled the Esk, the river takes several abrupt turns in a northerly direction and is joined by Meggat Water (8), with Stennies Water from the north of Westerkirk. Shut within a narrow vale, less than two miles broad, Esk flows circuitously and rapidly to Langholm town, where it is joined on the east by Ewes Water, and on the west by the Wauchope (5). The Ewes (10) from the north-east boundary flows through lovely defiles side by side with the highway from Carlisle to Edinburgh. Tending to the south-west, Esk descends through Canonbie, being swelled on the left bank by the impetuous Tarras (9), which rises at Hartsgarth. After passing through the wooded glen south of Langholm, past Canonbie village, Esk enters England, being joined below Nether Woodhouselee by the Liddel Water, and turns away westward to join the Eden and fall into the Solway. Liddel Water, after a long course through Roxburghshire, touches Dumfriesshire beneath Liddelbank and becomes the boundary between Canonbie and Cumberland for over five miles.

The historic Sark (13) rises in the Collin Hags, flows by Middlebie, between Half Morton, Kirkpatrick-Fleming,

Meeting of the Ewes and Esk, Langholm

and Gretna parishes and Cumberland, and falls into the
Esk south of Old Gretna.

Loch Skene is a mountain loch, four-fifths of a mile
long and less than one-fifth of a mile broad on the average,
lying 1700 feet above sea-level in a deep pot beneath Loch
Craig Head.   This still dark lake is the result of the

Loch Skene

confinement of the hill waters by moraines—relics of old
glacier movements—thrown in crescent shape across the
valley.   Out of Loch Skene issues the Tailburn, which
casts itself over a precipice forming the lovely cascade
known as The Grey Mare's Tail, altogether nearly 400
feet high.   The ruddy fleshed trout of Loch Skene is
much prized by anglers.

Around Lochmaben several lochs remain, namely the Castle Loch, 200 acres, Mill Loch, 70 acres, Kirk Loch, 60 acres, Hightae Loch and Marr Loch, 52 acres, and two "blind" lochs: Grummell Loch and Brummel, or Halleaths, Loch are now drained.

Loch Urr, 623 feet above sea-level, lies in Dunscore and Glencairn parishes, and partly in Balmaclellan. It measures 137·765 acres, of which 33 acres are in Balmaclellan, and 33 in Glencairn. Out of it flows the Urr Water. A few smaller lochs exist. Out of Townfoot Loch, Closeburn, flows a streamlet which descends through the wild ravine of Crichope into the Cample.

6. Geology and Soil.

The earth itself partly tells the story of its origin and growth below the surface and upon it, and of various forms of vegetable and animal life which have in succession existed in ages long gone by. In the crust of the earth —our only accessible book—we can read this ancient record from the minerals and rocks composing it. The term "rock" denotes either a material mass formed of one mineral or composed of more than one. Consequently it has been found necessary to classify the various rocks according to their characteristics and differences. These fall into three main divisions, with many subdivisions in each, namely (1) unstratified, massive, igneous, or eruptive; (2) stratified, sedimentary, or aqueous; (3) metamorphic or altered rocks. For convenience the terms igneous, aqueous, and metamorphic are generally used.

Grey Mare's Tail, Moffat

The igneous rocks are the earliest variety, lie at the very base of the crust, and have been pushed among and through the other rocks, somewhat like lava in its flow.

There they cooled slowly, solidified under pressure, and settled. Their composition and texture are very varied. Active volcanoes indicate what place and function this unseen basal foundation of the whole earth has. The aqueous, or sedimentary rocks, are found in layers— stratified; and because of this feature they are of the greatest importance to the inquirer, since they have been deposited at different times and within their composition preserve memorials of the various epochs they have passed through. Within their substance fragments of different minerals and rocks, of chemical precipitates, such as salts, and, still more instructive, of fossils of plants, fishes and animals, are found in well-defined groups. For example, in limestone caverns stalactites falling from the roof, and stalagmites rising from the floor, are formed by the deposition of carbonate of lime in solution; and the masses formed often have encrusted in them many foreign materials. Similarly crustacea, mollusca, zoophytes, and foraminifera ever eliminate carbonate of lime from water, and this on deposition makes solid rock. The ooze of the ocean bed filled with the skeletons of foraminifera becomes a kind of chalk. Coral, resembling limestone, is built up in similar manner. Nearly all the land surface and much of the sea-bed are of the sedimentary class. The winds too have been important factors in gathering sand and loose material together into situations favourable to stratification.

The third class of rocks is the metamorphic, consisting of the above two classes after they have undergone alteration. The action of irresistible subterranean forces, which, interacting with heat and water, pressed, crushed, and re-arranged the composition of the mineral elements giving them crystalline structure and properties, is seen in the resultant products. These are called schists and range from silky slates, or phyllites, up to coarse granite-like gneisses.

Thus the earth, upon examination, tells how it has been acted upon by fire, water, air, and life under law. Taking all these facts into consideration, geologists have laid down the order of succession of the stratified formation of the earth's crust as follows:

		Systems		General Characteristics
TERTIARY OR	CAINOZOIC	Pleistocene		Boulder Clay, Sands, Gravels.
		Pliocene		Shelly Sands and Gravels.
		Oligocene		Clays, Marls, Sands, Limestones.
		Eocene		Sands, Clays, Loam.
SECONDARY OR	MESOZOIC	Cretaceous		Chalks, Sandstones, Clays.
		Jurassic		Shales, Sandstones, Limestones.
		Triassic		Sandstones, Limestones, Gypsum, Salt.
PRIMARY OR PALAEOZOIC		Permian		Red Sandstones, Magnesian Limestones.
		Carboniferous		Sandstones, Shales, Limestones, Coal.
		Old Red Sandstone and Devonian		Red Sandstones, Slates, Limestones.
		Silurian	Ludlow Beds / Wenlock Beds / Llandovery Beds	Sandstones, Shales, Limestones.
		Ordovician or Lower Silurian	Caradoc Beds / Llandeilo Beds / Arenig Beds	Shales, Slates, Sandstones, Limestones.
		Cambrian		Sandstones, Slates, Limestones.

Dumfriesshire forms a part of a high Silurian table-land which stretches from Port Patrick across country to St Abb's head. This extensive belt consists almost wholly of massive grits, greywackes (whinstone), flags and shales, except in certain circumscribed areas where the carboniferous system is in evidence. This table-land is hilly, conformable to the system. The Silurian strata, however, having been subjected to great lateral pressure, bent, squeezed, inverted, and often times cleft, are found somewhat complicated, without any remarkable metamorphism. The hills, irregularly set, have taken shape and position, through no upheaval, but through the erosion of the valleys. They resisted the scooping forces which bared the hollows and sculptured the romantic craigs and dales. This Silurian area was in Palaeozoic times overlaid by a thick bed of Old Red Sandstone, which has totally disappeared through denudation, although it appears over the county boundary in Ayrshire and stretches away in a north-easterly direction to the Braid Hills, near Edinburgh. In the hollows worn out of this high table-land, first Old Red Sandstone, followed in order by the strata of the Carboniferous system, then rocks of the Permian and Triassic period were deposited to be in turn borne away except from some depressions in the vales of Nith, Annan and Esk. In Spango basin, on the northern boundary, granite, older than the Upper Old Red Sandstone, invades the Silurian system; and dykes of felsite, diorite and other igneous rocks appear in the Silurian area of the same age.

Before the Old Red Sandstone period passed away and as a result of volcanic action, igneous rocks appeared and in the form of slaggy, amygdaloidal andesites intervened between the sandstone and carboniferous strata.

The lava flow is traceable between Tarras and Birrens-wark, and sites of volcanic orifices are seen in Tarras, Liddel, and Ewes dales. Birrenswark Hill is simply a mass of lava standing up through a bed of Upper Old Red Sandstone. There is a visible outcrop of lavas and agglomerates (with dolerite and gabbro) on the northern boundary near Wanlockhead, Bailhill, Sanquhar, and Euchan Water; and volcanic vents of the Permian age can be traced in that district. Only volcanic tuffs are found in the Moffat area.

The Lower Silurian system, in its two-fold division of Llandeilo and Caradoc beds, is represented in seven groups of varying character and thickness. Examples of Llandeilo formation are seen near Raehills, of Caradoc at Hartfell Spa, and of Llandovery at Dobb's Linn, all near Moffat. Around Moffat there is a large zone of black shale accompanied with true deep-sea deposits, with grey shales and yellow shattery clayey bands. Between Moffat and Sanquhar lies the Dalveen group of hills, where a series of fine blue and grey greywackes and shales similar to those of the Llandeilo bed is found. The Hartfell black shales disappear to the west of San-quhar, their place being taken by grey sandy shales with dark seams through them; and in the district south of Eskdalemuir they pass into Wenlock beds. A line drawn from Ewes Water Head by Langholm to Mouswald forms the boundary between Llandovery and Wenlock. An-other line drawn from Langholm to Ruthwell indicates where the Upper Silurian is covered by the Upper Old Red Sandstone and by Carboniferous rock. An interesting portion of the Caradoc area—greywackes and grits with coarse conglomerates exposed—is found in the upper part of Penpont and Tynron, extending westward from Scaur, Chanlock and Shinnel Waters. These conglomerates contain pebbles of quartz, quartz rock, lydian-stone, blue and grey greywackes, grey shales and pieces of black shale. Graptolites were found among the fragments of black shale.

The Queensberry Group is a great series of massive grits of Tarannon Age, systematically jointed, with Barlae shales on top, and grey and blue shales at base resting on the lower black shale bed, and spreading over the south-western quarter of the shire into Kirkcudbright. In the Cluden Valley the black shale is thickly covered with drift, and just crops out north of Glenesslin. In the Scaur Valley there is a considerable area of the "Grieston" series, and in the blue shales of the vicinity the trilobite has been unearthed.

Of metamorphism in the Lower Silurian area there is a fine example at Bail Hill, Kirkconnel, where the transition from ordinary greywacke and shale into crystalline amorphous rocks is seen. The resultant mass is a curious, varied, undefinable blend. In the area around Wanlockhead the Lower Silurian rocks are traversed by two veins, one running north-west and south-east, the other west-north-west and east-south-east. These contain lead, iron, barytes, sulphuret of zinc, copper-pyrites, and silver. Gold is found in the alluvia of the streams. Both lead and silver are worked to profit. The lie of the strata in the lead-producing region is as follows: greywacke; Black

Jack (zinc-blende decomposing into clay, $\frac{1}{2}$ inch); "vein-stuff" (greywacke ground and mixed with quartz, $1\frac{1}{2}$ inch); calc-spar ($\frac{1}{2}$ to 1 inch); galena ($\frac{1}{2}$ inch); "vein-stuff" (quartz ore and quartz, 2 to 3 inches); blue greywacke (calcareous, $3\frac{1}{2}$ feet); hard quartz (iron pyrites in "flowers," 7 inches)—total, 8 feet 3 inches.

The Carboniferous system is well illustrated in several parts of the shire—in a tract from Langholm to Ruthwell, in Middle Nithsdale, and in Upper Nithsdale. The layers descend in the following order—sandstone, shales, reddened and yielding plants—coal measure. These Carboniferous

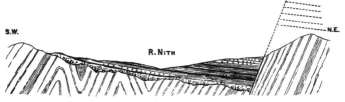

S.W.   R. NITH   N.E.

Sanquhar Coalfield

rocks still lie in ancient hollows worn out of the Silurian rocks in the Palaeozoic age, first filled with Old Red Sandstone, next eroded, and later filled with Carboniferous and Permian strata, which also were hollowed into the present depressions. This primeval Carboniferous valley —now called Nithsdale—begins at Kirkconnel, where a barrier would have diverted Nith into the Clyde. Consisting of sandstones—white, grey, and red—of dark shales with seams of coal, iron-stone, and limestone, this series is in evidence around Sanquhar, in Carron Water, and the Nith at Drumlanrig.

The Kirkconnel and Sanquhar coal-beds lie in an area measuring nine miles long with a breadth averaging from two to four miles, and is an extension of the Ayr coalfield. The strata descend 1200 feet, and the coal measures lie above the depressed surface of that part of the Silurian area which a fault has lowered. In different seams, with intervening strata, are found "creepie" coals, calmstone, twenty-inch, daugh, splint and swallow-craig coals. In one-half of the field, lying to the south-west, three doleritic dykes, throwing out intrusive sheets, disturb the measures and render the working of coal unprofitable there. In Upper Nithsdale the Silurian barrier did not sink beneath the sea-level till the latter part of the Carboniferous period. At Sanquhar red marls and clays, lying in the upper part of the coal measure, are available for the manufacturing of bricks, terra-cotta, pottery, tiles, etc.

Canonbie coalfield is said to represent the true coal measures of the central valley of Scotland. In the valleys of the Liddel and the Esk the nine following zones are represented in their ascending order : the Whita sandstone; the cement-stone group; Fell sandstones ; Glencartholm volcanic group; marine limestone group with coal seams; millstone grit; Rowanburn coal group; Byreburn coal group; red sandstones of Canonbie. In the Glencartholm volcanic zone, a number of new fishes, decapod crustaceans, phyllopods and scorpions have been found in the calcareous shale associated with the tuffs.

As Nith courses south to Drumlanrig the Carboniferous system, through which it has cut its channel, appears lying on the Lower Silurian rocks. Permian rocks appear above them both. The Carron Valley presents both Carboniferous and Permian rocks in varying positions, the latter sometimes overlapping the former. In the Thornhill Basin, that is from Blackwood north to the Lowther foot-hills, Carboniferous rocks of the earliest type appear, having fossiliferous red limestone bands lying above or below older and later sandstone, coloured red and reddish grey. The limestone is confined to the lower part of the basin, and is worked at Closeburn and across the Nith at Barjarg in Keir, where the Carboniferous rocks are best displayed. At Closeburn the strata are thus disposed: red shaly sandstone and purple and red mottled shales; red Magnesian limestone, 14 feet; red sandstones and clays, 18 feet; thick red almost pure limestone, 18 feet. Fossils common to the formation are found here. The other series of marine limestone found in Esk, Penton, Ecclefechan, and Kelhead have been wrought to commercial advantage. Kelhead is still an extensive work.

From Cumberland to Ayrshire the Permian rocks lie on the top of the Carboniferous system. They exist in two series—rocks of a lower volcanic character and a brick-red sandstone of considerable thickness. In the Permian age lava streams flowed over some parts of this area, then cooled into a band upon the Silurian or Carboniferous surface; and upon this porphyritic bed the stratified rocks of the Permian order rest. In places the porphyritic bed is interstratified with tuff and red sandstone: in others the brick-red sandstones are full of volcanic *débris* and bands of red volcanic tuff. Above Crichope

Crichope Linn

Linn, Thornhill, the Carboniferous rocks disappear, and porphyritic rock and sandstone of the Permian series take their place. At Locherben, between the systems, lies a bed of breccia made up of Silurian fragments, and blocks of porphyrite imbedded in a brick-red ashy paste. In the Dumfries Basin, where the Carboniferous rocks and volcanic series of Permian rocks are absent, the brick-red sandstones rest on a coarse breccia above the Silurian rock. This breccia is composed of pink felstone with hornblende, grey and purplish greywacke, grey shale, quartz rock and grey schist. There is a great demand for building purposes for the easily hewn Permian sandstones got in the quarries at Gatelawbridge, Closeburn, Locharbriggs, Corsehill, Annanlea, Cove, and Corncockle. In Corncockle Dr Duncan of Ruthwell discovered the footprints of extinct reptilia. The grey calcareous sandstone of King's Quarry and Auchennaight, Nithsdale, affords massive material for edifices such as Morton and Drumlanrig Castles. The Triassic rocks rest unconformably on all the older formations within Dumfriesshire.

The great ice-sea, in moving from the west in an easterly and south-easterly direction over this area, rounded off the hills and left their rock-faces in many parts scarred with the striae which tell of this passage. The ultimate result was the deposition of the boulder clays, beds and banks of gravels and sand in the valleys, the dropping of such a grand landmark as the historic "trysting" stone of Clochmaben in Gretna, and other transported boulders taken from Criffel across country over the Borders, the laying down of the fine glacial clay at Ryedale and

Hannahfield, Dumfries, and the formation of the vast moraines which are a striking feature at Loch Skene, and of the raised beaches seen near Queensberry.

The richest soils are to be found on the higher alluvial flats by the banks of the rivers and streams; but by Nith, Cluden, and Cairn, where there is a wide distribution of gravelly drift, the soils are generally light. As a general rule, the arable soil rests upon a subsoil of clay, especially in the more level tracts of country, in the valleys, or in the uplands. Nearly all broken land is tinged with a ruddy hue arising from the presence of oxide of iron. The Corncockle sandstone exhibits a light terra-cotta red colour and this hue is very prevalent in the sandstone belts upon the ploughed fields. The analysis of the stone shows —silica 95·10; alumina 1·16; oxide of iron ·85; lime ·31; potash 1·25; magnesia ·10; soda ·20; titanic oxide ·13; combined water ·90. The weight of the stone is 126 lbs. for every cubic foot.

7. Natural History.

In recent times—recent, that is, geologically—no sea separated Britain from the Continent. The present bed of the German Ocean was a low plain intersected by the waters of the present Rhine, which had among its tributaries the various eastward-flowing streams of Britain. At that period, then, the plants and the animals of our country were identical with those of Western Europe. But the ice age came and crushed out life in this region. In time, as the ice melted, the flora and fauna gradually returned, for the land-bridge still existed. Had it continued to exist, our plants and animals would have been the same as in Northern France and the Netherlands. But the sea drowned the land and cut off Britain from the Continent

Drumlanrig Castle

before all the species found a home here. Consequently, on the east of the North Sea all our mammals and reptiles, for example, are found along with many which are not indigenous to Britain. In Scotland, however, we are proud to possess in the red grouse a bird not belonging to the fauna of the Continent.

Dumfriesshire is a profitable resort for naturalists. Nearly all the trees and shrubs indigenous to Britain, 60 in number, are found flourishing within this area. There is no trace of the ancient Caledonian Forest, save what is preserved in the name of part of it—Holywood; but in remote glens there are patches of old self-sown, free growing copses of hazel and birch, with willow and alder in moist situations. The mosses sometimes give up bulky trunks of oak and Scots fir indicating how widespread the natural forest was. In 1756 the woods of Drumlanrig were felled by a storm and converted into rudimentary peat, a kind of transformation within historic times testified to by many peat hags. Some grand trees have survived storm and axe for centuries, and remain models for those landholders who have enriched the landscape with ornamental and commercial plantations, which have ripened and given place to other profitable woods. Nevertheless only one twenty-third part of the soil is thus occupied.

There yet survives round Drumlanrig Castle a small remnant of that

> " noble horde,
> A brotherhood of venerable trees,"

whose destruction by a Duke of Queensberry in the end of the eighteenth century stirred the poets Burns and Wordsworth into mournful verse. Some were so large as to produce the impression of an antiquity antecedent to the advent of the Douglases into Nithsdale in the fourteenth century. The policies of Drumlanrig contain a sycamore 100 feet high and 18 feet 3 inches in girth at 5 feet above ground, as well as an oak 89 feet high and 15 feet 10 inches in girth. Among other trees approaching 100 feet high are two silver firs respectively 13 feet 11 inches and 13 feet 4 inches in girth, and a Douglas pine 12 feet 4 inches in girth. With this destruction of the forest, shortly before the exit of the Douglases in 1810, is dated also the disappearance of the wild ox—

> "Mightiest of all the beasts of chase
> That roam in woody Caledon."

Few districts can show such immense beeches as those which encircle the lonely churchyard of Dalgarnock, several being over 4 feet in diameter. In Glencairn churchyard ash trees 12 feet in circumference, and oak 10 feet, are seen. On Craigdarroch estate, Glencairn, thick timber grows—beech, 16½ feet in circumference, at 5 feet from ground; horse-chestnut, 11 feet 10 inches; larch, 13½ feet, with clean stem 30 feet long, cut in 1906; silver fir, 14 feet 8 inches at base; sycamore, 14 feet. A crab-tree, 7 feet 3 inches, near Woodhouse, a horse-chestnut, 13 feet 6 inches, at Old Crawfordton, a sycamore, 14 feet 4 inches, at Castlehill, and yew trees over 10 feet in girth at Snade, are the finest trees in Cairn Valley. The juniper flourishes on the hill sides between Penpont and Moniaive. In Hoddom parish are several avenues of beeches of large proportions, referred to by Thomas Carlyle as "the kind beech-rows of Entepfuhl."

On the Springkell estate there are many magnificent forest trees, of which the following are examples: larch with girth of 9 feet 1 inch, at 5 feet, and 90 feet high; Norway spruce, 10 feet girth, 100 feet high; Scots firs,

10 feet girth and 60 feet high, and 9 feet girth and 80 feet high; silver fir, 17 feet 5 inches girth and 100 feet high; oak, 13 feet 2 inches girth and 69 feet high; beech, 13 feet 9 inches girth and 70 feet high; sycamore, 14 feet 9 inches girth and 60 feet high; Spanish chestnut, 14 feet girth and 55 feet high; lime, 13 feet 6 inches girth and

Hoddom: "The kind beech-rows"

70 feet high; ash, 10 feet 9 inches girth and 70 feet high; Huntingdon willow, 13 feet 9 inches girth and 60 feet high.

The trees around Closeburn Castle exhibit large proportions, an oak measuring 16 feet 8 inches in circumference, a beech 18 feet 1 inch, an ash in the Town Park 26 feet 8 inches, an ash at Shawsholm 19 feet 4 inches, and a Scots fir at Shawsholm, these measurements being taken at 2 feet above ground.

A silver fir, straight as an arrow, 105 feet high adorns the woods of Bonshaw, and stands beside a gigantic poplar rooted by the Kirtle. Similar fine examples stand on the banks of the Esk between Langholm and Canonbie.

Lochwood Oaks

The remnant of the old oak forest around Lochwood Tower, where gnarled, rugged, hoary specimens, many of them over 18 feet in circumference, still flourish, are probably the most interesting trees in the county. On either side of the ruined tower stand an equally massive ash and elm in vigorous life. These giants, the two great willows on the Cundy Road, Thornhill, the yews of

Drumlanrig and Snade, Dunscore, and many others equally noble, indicate how very suitable Dumfriesshire is for afforestation.

There is a wonderful variety of flowering plants and of grasses in this region, four hundred of the better known genera being represented between the sea-shore and the hill tops. The distribution of plant life is interesting. The ling (*Calluna vulgaris*), blaeberry, daisy, thyme, crowberry, thistle, geranium (*sylvaticum*), bugle, clover (*Trifolium repens*), watercress, dog-violet, wood-sage, male-fern and other well-known plants reach high altitudes. Foxglove and wild hyacinth are common up to 1100 feet. Ranunculus and Primula up to 600 feet—the peat-hag limit. Of Carex there are 35 varieties, of ferns 24, of Salix 16 (13 being in Nithsdale, 14 in Annandale, and 5 in Eskdale), of Potamogeton 16, of rush 10, of Poa 10, of Lycopodium 5, of Equisetum 6, and of Myosotis 5.

The following rare Flora flourishes—

(1) On the seashore: horned poppy, Isle of Man cabbage, sea-rocket, sea-holly, parsley dropwort, sawwort, sea-cat's tail grass.

(2) In holm-pastures and shady places: globe flower, monkshood, narrow-leaved bitter cress, dusky crane's-bill, dwarf whin, viscid Bartsia.

(3) In mosses and bogs: English sundew, marsh Andromeda, pale butterwort, prickly fen sedge, whorled caraway.

(4) On river sides and in water: great watercress, allwort, upright vetch, cowbane, quillwort, water-pepper.

(5) On hilly pastures and rocky places: vernal sand-
wort, alpine chickweed, alpine enchanter's nightshade,
cutleaved saxifrage, baldmoney, wintergreen (two varieties),
interrupted clubmoss.

(6) On high lands: alpine meadow rue, spring Poten-
tilla, alpine saxifrage, alpine Saussurea.

The following ferns are found:—Tunbridge filmy fern
(*Hymenophyllum unilaterale*), parsley fern, sea spleenwort,
green maidenhair, alpine woodsia, holly fern.

Of the vertebrates a goodly number survive. The
horns of a reindeer, with the bones of roe-deer, red-deer,
Bos primigenius, and the skull of the brown bear (*Ursus
arctos*), found under a peat-bog on the Shaw property,
indicate some of the animals formerly existing here.
The horns of a wild ox were found in Glencairn parish
in 1906: the skull of another, found in a Nithsdale moss,
is preserved in the Grierson Museum, Thornhill. Pen-
nant, in 1772, saw a herd of the wild cattle in Drumlanrig
Park. They were driven away about the end of the
eighteenth century. The red-deer became extinct locally
in 1815, as did also the roe-deer. The latter, however,
has been reintroduced for half a century and is increasing.
The fallow deer, where existing, is a domesticated animal.
The wild cat was last seen about 1813. The squirrel,
after becoming scarce, has begun to increase again. The
foumart, badger, and black rat are extinct. The weasel,
ermine, and vole still survive. The brown rat and mouse
are on the increase. The fox gives sport in Annandale
and Eskdale. The wolf was a wild beast of chase in
Eskdale in the thirteenth century. A small herd of wild

goats browses on Saddleyoke and Brandlaw in the wilds of Moffat. The otter still hunts and is occasionally hunted in the rivers. A seal occasionally comes up the Firth, where the porpoise and several species of whale are seen at times. The hedge-hog is common. The rabbit and European hare are subjects of commerce, and the mountain, or white hare, long confined to the high hills on the watershed, is now increasing on inland moorlands. The adder is less plentiful; the lizard is not uncommon. The glow-worm is common in Tynron, Moffat, and some other places.

Of the Avifauna a good report can be made, there being no fewer than 70 residential species, 31 summer visitants, 31 winter visitants, 30 occasional visitors, 56 species captured in the district, and 39 specimens not completely authenticated. To begin with the larger birds—it is about forty years since the last golden eagle was noticed in Nithsdale. About 1825 a very fine specimen was shot in Morton parish and long preserved in Burn farm. An osprey was supposed to build in Loch Skene, and a specimen was again seen there in 1881, while another was marked at Drumlanrig in 1840. The buzzard, or "gled," which formerly nested in the shire, increased there during the time of the vole plague, 1891–3, and now appears only in spring and winter. Peregrine falcons have from 12 to 16 nests; ravens the same number. The lesser hawks are common. The ptarmigan is extinct locally. The capercaillie is rare. Fourteen heronries and 145 rookeries still exist in the county. Rooks and jackdaws, despite the annual slaughter of them, are not decreasing. The chough is only an accidental visitor. Of birds which have been rarely seen in the shire are the great skua gull, little auk, red necked grebe, storm petrel, and the whiskered tern (*Hydrochelidon hybrida*). The bittern bleats but rarely now. The black-headed gull nests in 12 places and is on the increase. Since the

The Whiskered Tern
(*Shot near Friar's Carse, Holywood, in 1894, and now in the Royal Scottish Museum, Edinburgh*)

passing of "The Wild Birds Protection Acts," 1880–1904, and the County Protection Orders under section 76, "Local Government (Scotland) Act," 1889, there has been a stoppage of the cruel practice of destroying rare birds, and a consequent increase of such beautiful birds as the British jay, kingfisher, goldfinch, tufted duck, nightjar, cuckoo, owl, lapwing, and the smaller "minstrels of the grove." In summer the warbling of birds in the woodlands is a characteristic of the county.

The intelligent occupation of shootings has helped the increase of the more harmless members of the winged tribe, and the necessary preservation of the stock of the edible order. The shootings in Dumfriesshire, 144 in number, are now the basis of an important industry—that of game rearing and preserving. The pheasant, introduced some years before 1794, together with the increasing woodcock and snipe, have a prominent place in the woodlands. The red grouse (*Lagopus Scoticus*), the black grouse and his grey hen (*Tetrao tetrix*), and the partridge, notwithstanding their annual disasters, are plentiful. In 1869 no fewer than 247 blackcock fell in one royal shoot at Kirkconnel.

The lochs and sea-margins are vocal with the sharp screams of dotterel, sandpiper, goose, duck, gull, tern, and other waders, swimmers, and divers, of which there is a delightful variety.

Of the freshwater fishes of the Solway area, that is, from the Esk to Lochryan, there are 20 varieties. Nearly every stream contains trout; rivers have trout, sea-trout, grayling, and salmon. Nearly every loch has pike or "gedd." Roach is got in Annan and Lochar; chub in Annan; bream in Lochmaben lochs and Annan; smelt in Nith and Annan; and tench in Upper Nithsdale. There are also little goby, stickleback, flounder, minnow, loach, and Allis shad. Eels and lamprey are common. The sea lamprey comes up the rivers to spawn. The sturgeon is found in the estuaries. Salmon, shrimps, and oysters are a valuable asset.

The Loch Skene trout is a lovely fish with pink flesh and red spots, and weighs half a pound. The Castle Loch, Lochmaben, contains seven varieties of fish—pike, perch, roach, bream, chub, loch trout, and the famous vendace. The vendace (*Coregonus vandesius*), only to be caught by net, is an uncommon fish. It occurs in three lochs at Lochmaben, and in Windermere and Bassenthwaite lakes. It is said to be a ground feeder and shy of bait. Specimens weigh as much as one-sixth of a pound. When the vendace leaves its own element for the Annan river, it degenerates and soon succumbs.

8. Round the Coast.

The coast line beginning at Dumfries and continuing to a point east of Old Gretna, measures 32 miles. In its inward flow the tide touches the parishes of Gretna, Dornock, Annan, Cummertrees, Ruthwell, Caerlaverock, and Dumfries. The Nith, confined within its banks, flows in a stream 100 yards broad to Kingholm Quay on the left bank 2¼ miles from Dumfries. Here vessels of 300 tons discharge. Two miles farther down on the same side is Kelton, at one time a flourishing port where shipbuilding was carried on. Flowing between the green merse on the Kirkcudbright side and the opposite low lands of Caerlaverock, and leaving at low tide a sludgy margin, Nith reaches Glencaple Quay, less

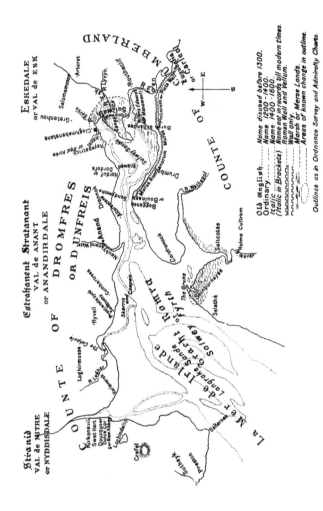

Map of the Solway

than six miles from Dumfries. Here there are 14 feet of water at high tide.

After this point the bank, cut by the river, begins to widen out and is known as Blackshaw Bank—six miles long from Nith to the channel of the Lochar and three miles broad. Thereafter the bank is called Priestside Bank and stretches other four miles to Powfoot. A good road runs parallel to the Nith, and passing the base of Banks Hill (297) and Plantation, leads to the magnificent ruined castle of Caerlaverock—the "Ellangowan" of Sir Walter Scott's novel, *Guy Mannering*. This moated stronghold of the Maxwells stands less than 50 feet above sea-level and is partly sheltered by a wood thrashed with sea winds and spindrift. South-east of the castle lies a small merse on the upper edge of which are found the Saltcot hills, reminiscent of the manufacture of salt here in the olden time. Writing in 1612, John Monipennie says, "Upon the banks of Sulway in June and July the country people gather up the sand within the flood marke, bringing it to land, and laying it in great heaps, thereafter they take the salt spring water and cast it upon the sand, with a certain device causing the water to run through the sand into a hollow purposely made to receive the water, which water being boyled in a little vessel of lead there is made thereof good whyte salt after the temperance of the weather." Between the Brow-well, with its sad memories of dying Burns, and Powfoot many of these pits are visible. Further east from Saltcots the dark stream of Lochar, after many a bend, touches the Solway just opposite the fine woods around the old castle and modern

mansion of Comlongon—a seat of the Earl of Mansfield—
and near to the Golf-links, one mile south of Ruthwell.
On this southern coast the territory does not rise 100 feet
above sea-level until the parish of Cummertrees is reached,
when a bank of that height is found just above Queensberry Bay and the village of Powfoot. Here the small
stream called the Pow burn falls into the Solway. Powfoot
is a growing fashionable watering-place of recent creation.
A few years ago, an insignificant hamlet of white-washed
houses, occupied by fishermen, less than a mile south-west
of Cummertrees village, it has been transformed into a
lively summer resort, where amid ornamental grounds,
or on greens for tennis and bowling, the bracing breezes
off the Firth may be enjoyed.

Thence for two miles inland the ground lies low and
flat, and at three miles from Powfoot the Waterfoot of
Annan is reached, after the traveller half-way has passed,
near the shore, Newbie Cottages, and thereafter the Mains
and ruined fortalice of Newbie. Newbie Tower was
formerly a seat of the Johnstones. A little distance west
of Newbie Cottages the boundary line between the Solway
Firth and the channel of the river Eden is placed. The
tide, however, goes 14 miles beyond this point up to
Alison's Bank Farm a mile distant from Gretna Green
village.

At Barnkirk Point, Annan Waterfoot, a lighthouse
stands, and from this point the river Annan takes almost
a semicircular course for nearly two miles up to Annan
town and quay. Steam vessels of over 500 tons can
sail up to this quay.

A magnificent railway viaduct spans the channel of the river Eden, and stretches from Seafield to a spot near Bowness on the English shore. One mile and a half east of the viaduct the Eden, near Dornockbrow, flows into two channels—Eden and Bowness Wath—and their waters meet again beneath the viaduct. The Dornock fishings were formerly very rich. On the green fields of

Solway Viaduct

Dornock, on 25th March, 1333, was fought a battle between Sir Antony Lucy and the Knight of Liddesdale, in which the latter was defeated and captured, many Scots also being killed.

The vast mud flats between the two kingdoms open out, and at a point south of Torduff the rivers Esk and Eden unite. Here the Esk becomes the national boundary, and its circuitous channel reaches to a point east of Red-kirk point, where formerly stood Rainpatrick Church. Here the waters of the Kirtle and the Sark, just united

Clochmabenstane

to the south of Old Gretna, flow into the Esk. Above this point Sark becomes the boundary.

On the flat land between the Kirtle and the Sark, at their junction, on the farm of Old Gretna, a few yards above highwater-mark, stands a historic stone, called Clochmabenstane, the last of a circle of stones which stood there. It is a granite boulder over 7 feet in height, and 17 feet in circumference. At it the wardens of the marches held their Courts in the olden time. Here was the termination of the Scottish ford at Sulwath—the ford by which the opposing armies often crossed ; and opposite to it, on the English shore, is Burgh-by-Sands, where King Edward I died. Around Clochmabenstane, in the battle of Sark, on 23rd October, 1448, Hugh Douglas defeated Percy with great slaughter.

The velocity with which the tide careers up Solway Firth with a high breast of waters is remarkable, and this characteristic inspired Sir Walter Scott to compare with it impulsive human affection : " Love flows like the Solway and ebbs like its tide." " In the winter season," writes Neilson, the historian of the Solway, " the scene, impressive under any conditions, is much intensified, especially if the tide is high and there is a southerly gale behind. Then the sea approaches with great speed, gaining as it goes : the wave is white with tumbling foam ; a great curve of broken surf follows in its wake : and the white horses of the Solway ride in to the end of their long gallop from the Irish sea with a deep and angry roar."

9. Coastal Gains and Losses.

The present configuration of the lowland portions of the shire, which have records for centuries of tillage on the same areas without perceptible change, leads to the conclusion that it is only under the surface on the one hand, or high up on the ridges on the other, we can find proof either of the loss or gain of land by the action of the sea. In the locality of Queensberry, on a hill behind Locherben, are observable three parallel lines of terraces at the height of 980, 1000, and 1100 feet respectively. The lower ledge, running from Garroch to Capel Burn, is broken at Capel and exhibits a base of stiff red boulder clay, overlaid with a stratum of fine sand, in turn covered with well water-worn gravel intermingled with sand. When the sea laved these high margins few eminences between Hutton and Corrie and the confines of Kirkcudbright emerged out of the waters covering Annandale, Nithsdale and Glencairn.

The sea on the west side has a barrier in Criffel and its spurs; but on the east side of the Nith a long ridge rising up from Dumfries to Trohoughton (312), an old fort and beacon, runs south and tapers out at the sea shore. Parallel to this ridge, another ridge four miles away descends to sea-level at Barnkirk Point. Between these parallels lies Lochar Moss, a little over 40 feet above sea-level. Tradition has preserved its suggestive history:

> " Once a wood and syne a sea;
> Now a moss and aye will be."

In 1754 Smeaton the engineer reported that Lochar could be reclaimed at a cost of £2952. At present the peat is being cut on Ironhirst Moss for commercial purposes.

The "finds" from strata in the moss, which is still insufficiently examined, do not contribute much of its story from the Neolithic age onwards. According to a writer in the *Statistical Account*, "There is a tradition universally credited that the tide flowed up this whole tract above the highest bridge in the neighbourhood. In the bottom of the moss sea-mud is found : and the banks are evidently composed of sea sand. A few years ago a canoe of considerable size and in perfect preservation, was found by a farmer when cutting peats, four or five feet below the surface. Near the same part of the moss, and about the same depth, a gentleman found a vessel of mixed metal...antiquities of various kinds are found in every part of this moss where peats are dug, even near its head, such as anchors, oars, etc.; so that there is no doubt of its having been navigable near a mile above the highest bridge, and fully twelve miles above the present flood mark." An extensive forest must have arisen on the area from which the sea receded, and stems and trunks of oaks which grew in the submarine clays are found, while further inland among the sandy and gravelly deposits the roots of fir trees are preserved in the moss.

On the west side of Nith a raised beach near Cargen Water, three miles south-west of Dumfries, and in Kirkcudbrightshire, was pierced for a well. After going through sand and silt for 15 feet, through peat for 18 feet, and through 14 feet of clay, gravel was struck. Into the clay, where marine shells were found, the roots of a fir tree had penetrated. Beside it were remains of charred wood, bundles of moss, and traces of phosphate of iron, as if indicating the work of man. The recession of the sea had thus permitted the growth of vegetation, and the woodland had been again submerged and covered with the later deposits. A 50 feet beach stretches from Park near Maxwelltown southwards to Cargenholm, and in the flats of Cargen a later beach of 25 feet in height indicates the recession of the sea. These have their corresponding levels on the banks of Mouswald and Ruthwell. But utilitarian gains of merseland have been small.

At its lower part the Lochar Water was confined within constructed embankments which on giving way permitted the river to overflow lands already reclaimed and to make a new channel through the sand and silt. At the head of the estuary important changes have taken place in historical times. A carefully prepared map dating from 1552 shows what these changes have been, and how the Solway has been receding, leaving inland many broad acres now under the plough. The high contour lines around Gretna indicate how strong have been the efforts of the tide to wash away that solid barrier under which the rivers have laid their ever increasing deposits forming a larger delta. This map shows with great precision above Redkirk Point (25 feet) a sketch of the handsome structure of a church, with tower and steeple, called Rainpatrick, then existing, which has been entirely washed away together with the foreland by the sea. At this place the monks of Holme Cultram beyond Solway possessed saltworks which were ultimately acquired by the monks of Melrose in 1294.

In living memory the Eden and the Solway tide have encroached on land between Seafield and Dornockbrow— the latter name indicating a slight elevation—and removed a " merse " from 60 to 80 feet broad. Barricades erected to stop these encroachments were swept away by high tides and the sea remained conqueror. A contemporary chronicler records in November, 1627, the excessive rise of the Solway tide, which surrounded " the house of Old-Cock-pool," carried off seventeen of the salt makers on Ruthwell Sands, and destroyed many cattle.

10. Climate and Rainfall.

The position of Dumfriesshire, surrounded on three sides by high hills, and on the fourth side lapped by the warm western seas, tends to the creation of a climate on the whole mild and productive of fertility and longevity. The force of wind and saline rain, driven from the ocean into the valleys, is moderated by the interposition of many high barriers which affect atmospheric conditions, so that sunshine, shower, mist, and drying winds have a relative distribution, in consequence of which the climate is temperate as to heat and rain. Within the county there are twenty-four meteorological stations, including the national one at Eskdalemuir. The names and positions of the three affording statistics given here are Drumlanrig

Gardens in Middle Nithsdale, 191 feet above sea-level;
Dumfries, 155 feet; Comlongon in Ruthwell, 74 feet.
To these may be added Cargen, 85 feet, on the Kirkcudbright side of the Nith below Dumfries. Thus every
variation can be appraised.

The average barometer for the year 1909 taken at
the Crichton Institution, Dumfries, above the tidal limit,
was 29·818°; but there was a higher reading at Cargen,
and a lower at wooded Drumlanrig. For the past fifty
years the average barometer recorded at Cargen indicated
29·827°, and that is the mean of all the records, between
the highest, 30·928°, which occurred on 9th January,
1896, and the lowest, 27·620° which was read on 8th
December, 1886.

The temperature recorded for 1909 was on the general
average 46·2° Fahr.; but at Cargen a little higher, 46·5°;
and on an average of 50 years there 47·6° or 1·4° higher.
The last highest records are 82° at Drumlanrig on 13th
July, 1909, and 87° at Jardington on the preceding day;
and the lowest 1·3° at Eskdalemuir on 27th January.
The highest temperature for 50 years was 90·4°, taken
at Cargen in August, 1876, and the lowest taken there
in December, 1860, was − 4°. The highest annual average
temperature was 49·2° in 1868 with the barometer at
29·758°; and the lowest, 45·3°, in 1892 with the barometer
at 29·811°. Snow fell on 19 days during 1909. The
temperature of the Solway Firth, having the benefit of
the "Atlantic Drift" of heated surface water, is higher
than that of the rivers falling into it, and its water has
a higher temperature than the air.

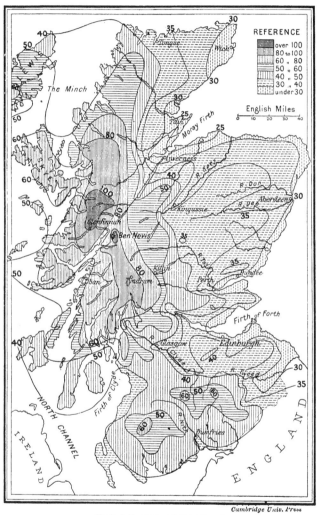

Rainfall map of Scotland
(*By Andrew Watt, M.A.*)

The last general average rainfall was 49·91 inches, with slightly over 50 inches at Cargen. Here in 1909 the fall was nearly 3 inches over that average, and nearly 8½ inches over the average of 50 years, which was 44·18. The wettest year, 1872, showed 63·50 inches, and the driest, 1880, only 30·77 inches at Cargen. In 1909 the wettest areas in the county were around Jarbruck, Glencairn, 59·42; Kinnelhead, Beattock, 59·20; Langholm (Westwater), 55·23; Moffat (Craigielands), 53·73; Ewes, 54·30; Cargen, 52·65; Eskdalemuir, 51·20; Drumlanrig, 47·99. There are dry spots around Dumfries—Joybank, 41·17; Comlongon, 41·68; Crichton Institution, 43·55. Stations at Canonbie, Lochmaben and Lockerbie showed over 45 inches; but two other stations at Lochmaben indicated 43·91 or an excess of 3·2 upon an average of 18 years. The average rainfall at 73 stations in Scotland was 41·36, so that, except in Dumfries itself, the average for the county is above the general average of the kingdom.

Dumfriesshire enjoys about one-third of the possible sunshine. During 1910 the sunshine recorded at Comlongon, Ruthwell, was 1444 hours, at Dumfries, 1368, and at Eskdalemuir, 1275. The highest record was 223 in May at Comlongon, and the lowest 31 in December at Dumfries. The average per month was 120 at Comlongon, 114 at Dumfries, and 106 at Eskdalemuir. From March to September, the average was above 152, computed from the three registers.

The humidity of the shire is 86 p.c. While at some stations the weather is not registered as calm or variable, at others it is proved to be very variable. Few thunder-storms occur. Twenty-five per cent. of days are overcast.

The prevailing winds are out of the west quarter, the north-west wind however blowing, at one station, one-third more days than the south-west wind, which shows a considerable shifting towards due west, where the lofty Galloway ranges cause deviations. According to observations made at Cargen, which has its own peculiar wind record, in 1909 the west wind blew 108 days, in 1908, 129 days, and in 1907, 121 days, but in 1906 only 96 days. The variations noted at Drumlanrig, Dumfries, and Cargen are remarkable. Easterly wind is not prominent. Take for example the differences of direction of wind shown at three places contemporaneously in January, 1909, reckoning from north round by east and south to north again:

Drumlanrig, north, 0, 0, 0, 2, 0, 24 west, 2, 34 north-west.
Dumfries, north, 6, 1, 3, 7, 3, 20 ,, 12, 8 north-west.
Cargen, north, 10, 0, 0, 0, 1, 6 ,, 28, 17 north-west.

This proves how surroundings influence climatic effects. Drumlanrig has an immunity from thunderstorms and gales in comparison with Dumfries and Cargen. At the latter station the mean wind-force is sometimes higher, sometimes lower than at Dumfries, although it is in no place excessive, being seldom scaled above 3. The gales which devastated Scotland in 1879 and 1883 spent their full force on the woods of the county and levelled hundreds of thousands of splendid trees.

The conjoint interaction of sunshine, rain, mist and breeze makes the climate agreeable and invigorating, while the facility for artificial drainage is another factor in the increasing immunity of the inhabitants from diseases which were formerly a sore scourge. Everywhere there is a conspicuous absence of dust in the atmosphere, there being few public works to pollute it with smoke and less than usual road dust swirled up from the tough highways, which are made of the best kind of macadam—felstone dolerite and the hardest greywacke.

11. People—Race, Type, Language, Population.

Many races have contributed their blood to form the native population on the soil of Dumfries; and, in consequence, although many families bear ancient names, and trace their origin to a remote antiquity, there is no distinctive type of individuals left. Of the aborigines, of an Iberian type, of which the Gaway or Galwegian Picts were probably the expiring remnant, there are so few traces that conjecture takes the place of record. In that part of Northern Britain to which early geographers assigned the honourable place-name of Valentia in which Dumfriesshire was included, the Romans found the folk called Selgovae, probably hunters, keeping to the east side of Novios or Nith, and the Novantae, also known as Niduari and Picts, upon the Nith and westward from it. On the south side of the Solway they located Brigantes, who occupied part of the province of Maxima Caesariensis. The Brigantes, evidently a Brythonic—

British or Welsh—race of fighters, overran their neigh-
bours, and finally cornered the painted pagans in the
wilds of Galloway. In Galloway these Picts remained
distinct, with separate laws and customs, probably too
with language modified after their contact with Celts
and Iro-Northmen, an almost independent race, who
fought at the Battle of the Standard, and a strong political
factor down to the fourteenth century. This forced
seclusion may account for the greater abundance of
primitive remains, weapons, and dwellings of the Neolithic age in the district west of the Nith. It was among
these pagans, whose thieving habits secured for them the
opprobrious titles of "Galuvet" and "the wild men of
Galloway," that the British missionary, Ninian, laboured.
Of their tongue we appear to have survivals in the many
unexplained monosyllabic place-names descriptive of certain
permanent landmarks in the shire.

After the removal of the Roman arms, the British,
or Kymry (hence Cumbria, Cumberland, Cummertrees)
spread over Southern Caledonia, and were piously ministered
to by British missionaries whose churches were dedicated
to Martin, Ninian, Blaan, and Mungo—foundations still
existing. Holding their lands in Dunscore from time
immemorial, the family of Welsh, the last of whom there
was Jane Welsh Carlyle who possessed Craigenputtock,
may well have sprung from that early British race. Their
language survives everywhere in relation to prominent
objects described by *caer, lin, pen, cors, craig, alt, man, tre,
ros* and other terms. Their "caer," or fort, is much in
evidence. Many families, long bound to the soil here,

and named Karrs or Kerrs, Carsons, Carruthers, Carlyles or Carls, Crechtons, Kirkpatricks, locally known as Caerpatricks, may be the descendants of the early holders of the forts.

In turn the Briton or Welshman was overrun by two streams of Celts, one coming direct from Ireland and

Craigenputtock: Carlyle's House

another circuitously descending out of north-west Strathclyde. It may be suspected that they found traces of a primitive people like themselves, with a similar if not an identical tongue. Their teachers brought the cult of Patrick, Bridget and Michael to churches named after these patron saints. If the conquest of the Gael was not complete their fixture of place-names was, so that this area might still find a place in the Highlands, so prevalent are the *auchens, bens, glens, duns, drums, bracks, minnys* and *corries*. Families too are proud of old, not imported, clan-names—for the clan system survived to the fifteenth century. MacRory, MacMath, MacCubbin, Mac-Gowan, MacCririck, MacDouall, MacDougal (Dubhgall), MacLachlan (Lochlan), MacCulloch, MacMurdo, Ma-gachan, MacEwan, Kellock (Ceallach), MacMichael, MacMillan, Macglethery (i.e. MacGregor) are local names of people, some of whom had chieftains, or feudal lords to own and protect.

A charter of the twelfth century, given by Radnulf, son of Dunegal, at " Dronfres," was witnessed by " Gil-christ, son of Brun, Glendonrut Bretnach, Gilcomgal Macgilblaan, Udard, son of Uttu, Walder son of Gil-christ." Here we see the vassals or friends of Nithsdale's great over-lord drawn from all the races, from the ab-original to the Teuton and Norman. " Gilcomgal Mac-gil-blaan," that is, the servant of the monastery—the son of the servant of (Saint) Blaan, is a most interesting reminiscence of the Celtic monastic cell of Saint Blaan in Caerlaverock parish, and a reminder of an ecclesiastical descent. Brun, Blain, and Gilchrist are still family names in the shire.

Thus the fusions went on. For a time Dumfriesians, like the Swiss of some cantons, may have been trilingual. The mixed population next met invading English, Danish, and Viking Northmen. A notable instance of the heroic racial spirit is recorded regarding a British leader, named

5

Constantine, who fell at Lochmaben, in 879, in a gallant attempt to lead his countrymen to the aid of the Motherland in Wales. The earlier Scandinavian emigrants, who crossed the seas to settle in eastern England, in course of time found their way across Northumbria to the Celtic region of Dumfries, where they left traces in numerous place-names, and, probably later, in the runes on Ruthwell Cross. Everywhere one finds localities distinguished by Anglo-Danish words bearing on their composition— *-dale*, *-garth*, *-wald*, *-myre*, *-haugh*, *-cleuch*, *-shaw*, *-burn*, *-water*, *-holm* and many others.

The coming of the Northmen round the west coasts, in viking expeditions, finally to settle in Cumberland, Dumfries and Galloway as farmers, had a modifying effect upon the tongue of the earlier Scandinavian and Danish conquerors, of which there are manifest traces on both sides of the Borders. The residence of the Northman is remembered in Arkland (*ergh*, a sheiling), Hartfell (*fell*, a hill), Waterbeck (*beck*, a stream), South Grain Pike (*grain*, a tributary brook), Middlebie and Lockerbie (*bie*, a farm), Murraythwaite (*thwaite*, a clearing, or old pasture land), Burnside and Burnhead (*sætr*, a pasture, and *hefd*, a head), The Rigg, The Riddings, and many another place. Indeed when the modern farmers in these places speak of the elding (fuel), rice-fence (*hrís*), the sile or siler (*síli*, a sieve), the handsel (*handsöl*, a bargain), a gowpen-full (*gaupn*, both hands full), a quey (*kvíga*, a young cow); and when the poachers of the *hopes*, *sikes* and *gills* of Eskdale refer to the "leister" (*ljostr*) or "waster," with its witters—barbs—and the "roughies," or dry branches which light "the burned water," they are using the nomenclature of their predecessors there—the North-men, farmers and hunters. Of a Frisian settlement in Dumfries, which by some is supposed to mean "the fort of the Frisians," there is no trace, and no record. To what extent the residence of the Northmen and the infusion of Norse blood changed the native population and the vernacular we have little evidence. One church, at least, bears a Northman's name, Closeburn (Kil, Osbiorn). The influence of the Anglian may be trace-able in that similarity to modern German pronunciation, both consonantal and vocalic, which distinguishes the Dumfriesian's expressions, and also in the longer persist-ence in the vernacular of words which come with greater fluency from Border lips, such as, *yird*, *nowt*, *gang*, *gate*, *byre*, *muck*, *scart*, *scoot* and others equally definitive.

The introduction of Anglo-Norman settlers in Cum-bria by Prince David for a time affected local blood and speech to a slight degree. Of the old family names—Mundells, Mounceys, Menzies, Bruns, Herries—few survive. A century ago, in the old English tongue, which the rural Dumfriesian spoke, there were survivals from the Anglo-Normans in terms in common use, such as *dule*, *gabbe*, *poke*, *pooche*, *tache*, *palmie*, *sort*, and the like, which were just as likely to have survived from feudal times as to have been imported in the era of the later Stewart monarchs.

The development of railway travelling, by which the people can so easily shift their habitations, has markedly affected the stratification of the languages preserved in the

Lockerbie from Mains Hill

local vernacular, and even the broader accent is on the wane. Still on the Borders there is a dialect somewhat different from those observed in Cumberland and Northumberland. In Annandale, words ending in *ee* or *ea* are pronounced *ei*, or *eye-ee* : *oo* is pronounced *ow* : *ue* or *ew* becomes *ĕ-yu*; true being pronounced *trĕ-yū*. The personal pronoun *I* is pronounced *aw*. *I am going* becomes *Am gawn*. *I'll make sure*, becomes *A's mak sure*, as anciently it was *A's*, or *I's*, *mak siccar*. Short *o* is generally pronounced long, so that a *rod* becomes a *road*, and *vice versa*. The letter *a* is often shortened to permit a more liquid sound to follow the letter *l*, thus *awl* becomes *ä-ll*. A curious iteration is noticeable in Hoddom, where the words, *give it to me*, become *gae me tit*. In Lockerbie a couplet runs:

"Yow an' meyee, an' the bern dor keyee,
The sow an' the threyee weyee pigs."

In Canonbie the pronoun *thou* is pronounced *ta*; *is* is used for both *art* and *are*. "Is ta gawn tae Kerle?" (Carlisle) is answered affirmatively "I is."

The peculiarities in the dialect of Nithsdale are fast dying out. One still hears the following exchanges: *A'm* (I am), *ee'r* (you are, thou art), *hay's* (he eyees) (he is), *oo'r* (we are), *y'ir* and *ee'r* (you are), *thä'r* (they are), *wäs* (was), *wŭr* (were), *hes* (has), *micht* (might), *cŭd* (could), *wŭd* or *wad* (would), *sŭd* (should), *wull* (will), *man* (must), *-in* and *-an* (-ing), *ŭ* for *i* (*budden* for bidden). A good specimen of the Nithsdale *patois* is the following : "Can ye gie's a wee puckle o' 'oo to stap i' the nebs o' ma shoon, for they're unco shauchly and they'll coup me

owre i' the glaur ? "—" Can you give me a small particle
of wool to stuff in the points of my shoes, because they
are very capsizeable and they will turn me over in the
mud ?"

In 1901 the population of Scotland was 4,472,043,
of whom 72,571, or one sixty-seventh part, resided in
Dumfriesshire, there being 34,362 males and 38,209
females, or 67 persons to every square mile. Dumfriesshire was thus fourteenth of the counties for numbers
of inhabitants. In 1911 the population of Scotland had
increased to 4,759,445 persons, and that of this county
to 72,824 persons. The inhabitants of the county now
form one sixty-fifth part of the entire population, and in
regard to their other relations stand precisely the same as
before. In 1881 the population of the county was 76,140;
in 1851, 78,057; and in 1801, 54,597.

12. Agriculture.

After the cessation of the wars with England, the
Scottish Borderers devoted themselves assiduously to
tillage and the rearing of cattle and sheep. Great
impetus was given to agriculture in the end of the
eighteenth century through new methods introduced by
Sir James Kirkpatrick of Closeburn, and his successors
the Stuart-Menteiths, by Douglas of Kelhead, and
Patrick Miller of Dalswinton and others, who converted
poor lands into show estates by improved drainage and
the advantageous use of lime. In Middle Nithsdale farmers

secured a good financial position on the transference of the Queensberry estates to the House of Buccleuch, when the latter for compensation set aside the long leases granted by the Douglases, and in years of great commercial prosperity beautified and improved all the holdings. The troublesome system of commonty in course of time died out.

The present importance of the farming industry may be appraised on consideration of the fact that there are 2692 holdings upon an available land surface of 664,119 acres, of which only 30,275 acres are occupied by woods. Besides these acres 26,175 acres of foreshore and tidal water remain. Of the 2692 holdings, 497 are under 5 acres, 882 under 50 acres, 1146 under 300 acres, and 167 over 300 acres. The land is thus portioned out: 382,112 acres of mountain and heath-land used for grazing; 118,733 acres of permanent grass; 134,486 acres of arable soil; 30,275 acres of woodland. The valuation of the county is £450,062, exclusive of railways £42,546.

Tenants occupy 223,710 acres laid down in grass and crops, and owners retain 29,509 acres. Of these fields 39,349 acres produce 20,054 tons of hay from permanent grass, and 17,715 tons from clover, sanfoin and other sown grasses. Oats cover 41,333 acres and yield 189,299 quarters, or an average of 36·64 bushels to the acre; barley takes up 590 acres, and yields 2751 quarters, or an average of 37·31; wheat is grown on 65 acres, and produces 320 quarters, or 39 on an average, commanding a good price. Rye is grown on 17 acres.

Large tracts are required for turnips and swedes, 17,280 acres yielding 305,931 tons of these roots; 3450 acres produce 25,543 tons of potatoes; and 305 acres produce 4824 tons of mangold.

Among other products are beans 14 acres, peas 4, carrots 34, onions 1, cabbage 226, and rape 304 acres. No flax is grown. There are many sheltered and sunny patches suitable for fruit growing, but this industry has not become popular. Apples grow on $21\frac{3}{4}$ acres, pears on $2\frac{1}{4}$, cherries on $1\frac{3}{4}$, plums on $2\frac{1}{4}$, and there are $38\frac{1}{4}$ of mixed orchards. Strawberries are grown on 28 acres, raspberries on $5\frac{3}{4}$, currants and gooseberries on $20\frac{1}{2}$, and there are $23\frac{1}{4}$ acres of mixed small fruit.

The work of this vast region employs 5878 horses including brood mares, and 1868 young horses await breaking in. Of 62,359 cattle upon the fields, 21,628 are cows and heifers, and 40,731 are calves and older cattle. 556,988 is the grand total of the fleecy flock, of which 243,369 are ewes kept for stock, and 227,506 lambs. Of 8546 pigs, 1096 are breeding sows. Of the 6517 men and 1393 women employed in agriculture in 1901, 3644 were farm servants. The latter number was a decrease of 1211 men since 1881.

The distribution of the various breeds of sheep in the county is interesting, 140,000 cheviots being kept for breeding on the green hills of the districts round Langholm, Lockerbie, Moffat, Closeburn, and Sanquhar. Blackfaced sheep to the number of 100,000 are kept on the high uplands and heathery hills of the north and west. Small flocks of Leicesters and Yorkshires, numbering

about 2000, graze on farms around Dumfries, Lockerbie
and Moffat. Crossbreds, Oxfords, Suffolks and Shrop-
shires, 2000 in all, are seen on good farms near the
Solway. The South Down, Dorset and Spanish breeds
are also represented in small numbers. In all 244,000
ewes form the breeding stock of the sheep-masters in
Dumfries.

The cattle trade is to the farmer a matter of first
importance, and sales of live-stock are periodically held in
Dumfries, Annan, Langholm, Lockerbie, Thornhill and
Carlisle. The following are the numbers annually disposed of in the auction marts of one firm : Langholm
(fat and store stock), cattle 600, sheep 12,000 ; Dumfries
(fat stock), cattle 2500, sheep 25,000 ; Lockerbie (fat and
store stock), cattle 2700, sheep 120,000. In addition
1100 pigs and 1500 calves are sold. At Annan, the
numbers of live stock sold are increasing year by year.
A speciality is the sale of short-horn bulls, 900 being
disposed of at special spring auctions. Many store cattle
are sold in May and June, and 10,000 lambs after
Lammas.

At the mart, Thornhill station, during the past five
years, the average sales have been 1278 cattle, 41,029
sheep, 64 calves, and 91 pigs, of the value of £43,239
annually. In 1907 nearly £50,000 worth of live-stock
was sold here.

The White Sands of Dumfries have for generations
been famous for the sales of horses and other live-stock.
Seed markets are also held in Lockerbie, Annan and
Dumfries in the spring.

Cattle Fair, Dumfries

The 30,275 acres devoted to timber growing produce a considerable tonnage of both soft and hard wood, which is required for props in pits, fences, cart-building, and various other industries.

13. Industries and Manufactures.

The volume and value of the industries, trades, and manufactures of the county may be calculated from the number of persons employed in the various branches of activity. In 1901 agriculture employed 7910 workers and domestic service required 4935 persons of whom 4053 were women. Building required 2158 hands, besides workers in quarries 1412, 213 men preparing wood, and 90 brick makers. The latter number includes tile makers. Metal workers numbered 1182, and workers in precious metals 115.

The textile industries employed 885 men and 1065 women : 892 men and 1488 women were engaged in making into dress these as well as imported goods, while 239 drapers sold them. Commerce connected with this and other industries was carried on by 585 persons who required 2128 carriers and transport agents. Skinners and tanners numbered 116 ; stationers and other workers with paper were 192 ; workers in chemicals numbered 75, and there were 1637 general dealers. No fewer than 1990 men and women were engaged catering for the others ; and 1220 professional men and women—teachers, doctors, lawyers etc., and 359 public servants looked after the education, health, and peace of the community. Over 5000 males and 20,000 females of working age had no specified employment.

The great sales of sheep and cattle at Dumfries, Annan, Lockerbie, Langholm and Thornhill, and the disposal of the various cereals, are productive of work and wealth. Nurseries for trees, shrubs, and flowers around Dumfries and Annan afford labour for many hands. The shire is not an industrial and manufacturing centre, although certain industries continue to thrive in it. The extensive sandstone quarries of Gatelawbridge, Closeburn, Locharbriggs, Annanlea, Cove, Corncockle, Corsehill and others, which formerly had a large export trade, have recently decreased their output; and the limestone quarries at Closeburn, Barjarg and Kelhead have not been so active as formerly. The mining industry in Kirkconnel is increasing. Lead working in Wanlockhead still flourishes.

A new industry has been recently started on Ironhirst Moss, part of the great Lochar Moss belt, for the utilisation of the peat, which is converted into compressed peat blocks and into sulphate of ammonia, under new processes.

Large numbers of the inhabitants of Dumfries town are engaged in various mills and works on both sides of the river, wherein cereals are ground, and yarns spun and woven into tweeds, gloves, hosiery and other articles for wearing. Dumfries tweeds, gloves, and silk underclothing have a high reputation in the market, and demand for them is on the increase. Tan works still exist. Forges for the production of agricultural implements and engineering shops employ considerable numbers of mechanics. To carriage building, for which the town was well known, motor-building has been added; while the making of jams and confectionery, dyeing and laundrying give employment to many hands. Being the chief emporium for a large agricultural area, Dumfries is busy in the transaction of all kinds of commerce, and is abundantly supplied with merchants in places of business supplying the needs of a practical community. The shipping industry in the Nith has become attenuated.

Annan, near the centre of a large agricultural district, is similarly busy, having engineering works, the famous mills for making "Provost Oats," some shipping trade, distilling "Johnie Walker" whiskey in the Annandale distillery, brickworks, nurseries, fisheries, and cattle markets.

In Eaglesfield village 70 persons find employment in tailoring. Langholm is noted for its six tweed-mills, wherein 660 male and female workers are engaged. A tannery employs 30 hands. Glen Tarras Distillery and Langholm Distillery, at present not working, afforded labour for others in Langholm.

At Sanquhar, in place of the thriving weaving industry which has disappeared, there are terra-cotta brick working, brick and tile making, coal-mining, a little sandstone quarrying and laundry work—all, however, on an inconsiderable scale.

A most important industry has sprung up in recent years in Thornhill—the bacon factory, giving employment to 50 hands. The latest modern plant has been set down to overtake the conversion of above 15,000 pigs per annum into bacon, sausages, cooked hams, and all kinds of table delicacies such as glass-meats, braised tongues, pork-pies and other tid-bits. Under steam pressure the bones are made to yield edible fat, and the residuum is used for manure. There is also a bacon curing establishment in Sanquhar.

The amount of printing done within the shire is very considerable in the production of books, the local newspapers, and in the execution of contracts from without.

Large numbers of men and boys are employed in the extensive woodlands of the county, both planting and cutting down the ripened trees. Several estate saw-mills are employed for local requirements, but large quantities of hard and soft timber are exported to mining centres and to districts manufacturing articles out of wood.

In 1897 the Charity Organisation Society of Glasgow leased Mid Locharwoods, a farm of nearly 500 acres in Ruthwell parish, as a labour colony for providing work for the destitute unemployed, and for the reclaiming of Lochar Moss. The number of employees fluctuates. The directors have recently purchased the farm.

14. Mines and Minerals.

Coal, lead ore, zinc ore, silver, limestone, igneous rocks and clay are mined and quarried in the county. The output, however, is limited. Coal is worked in two areas only—in Upper Nithsdale at Sanquhar and

Kirkconnel, and in Canonbie at the Old Colliery and Blinkbonny. The Sanquhar-Kirkconnel coalfield (which, as the chapter on Geology indicates, is a continuation of the Ayrshire field) with its three pits employs (1911) about 850 men and boys, raising about 300,000 tons of coal annually. A recent return (1907) showed an output of 245,624 tons, valued at £99,273, or 8s. 1d. per ton, raised by 437 hands under ground and 121 above, in all 558. The recent sinking of new shafts in Kirkconnel has increased these numbers. The latest return (1910) shows an output of 279,912 tons of coal, of value £87,473, or 6s. 3d. per ton, raised by 811 hands of whom 676 were employed in Dumfriesshire. The return includes the work of 135 men in the small Argyll coalfield.

The Canonbie coalfield is small but rich, and produces 18,000 tons annually, 105 men and boys finding employment in it.

Only one metalliferous mine is at work in the county, that at Wanlockhead, belonging to the Duke of Buccleuch and adjoining the well-known mines at Leadhill. Here 106 persons are employed above ground and 47 below. The output for 1909 was 1783 tons of lead ore and 107 tons of zinc ore. The output for 1910 was 2500 tons of lead ore, and the number of men employed 250. One hundred years ago 118 men were employed, producing 874 tons of lead, worth £5000.

Silver to the extent of 12,500 ounces was extracted in 1910 : in 1909, 8915 ounces. The total value of the dressed lead ore and silver in 1909 was £22,000.

An antimony mine was opened in 1760 at Glendinning,

Westerkirk, and before 1798 had produced 100 tons of the regulus of antimony worth £8400. Working stopped about 20 years ago.

There are 22 quarries employing 224 inside and 362 outside, and their output is as follows : igneous rock, 8156 ; clay, marl, brick-earth, shale, 7781 ; granite, 2692 ; limestone, 17,096 ; sandstone, 90,104 ; in all 125,829 tons. The values of these products were : igneous rock, including granite, £1274 ; clay, £341 ; limestone, £1109 ; sandstone, £38,868. This shows a considerable decrease from the output of former years. The new-red-sandstone quarries of the county are famous for stone suitable and easily wrought for building purposes and monuments locally, and in request in cities on account of weathering qualities when in contact with smoke and other deleterious elements in the atmosphere. The most notable of these quarries are Closeburn, Gatelawbridge, Locharbriggs, Corsehill, Cove and Corncockle.

The extensive new-red-sandstone quarries in Closeburn parish, which in 1903 produced 68,000 tons of excellent building stone, suitable for sculpture with a fine finish, and employed 250 men, in 1910 put out 21,000 tons at the hands of 60 men. The extensive lime quarries at Kelhead, Cummertrees, at present employ 19 men with an output of about 4000 tons of lime.

No coal is exported by ship from Dumfriesshire, but 4098 tons of coke are imported coastwise. 6094 tons of coke of the value of £3253 are manufactured in the shire.

15. Fisheries.

The Solway Firth from time immemorial has been famous for its fish, which remunerated the monks of Melrose and of Holme Cultram, and graced the royal table from the time of the Bruce onward. The Solway fisheries have often occupied the attention of the legislature and High Courts, while several Royal Commissions have investigated the area, statutes, fish, nets, and other matters relative to the Solway.

There are two distinct classes of fishing in Solway— the white and the red (or salmon). The white fish taken are cod, flounders, herring ; the red are salmon, trout, herling, sparling, and shrimps. The interests of the two classes of fishers often conflict on account of the ancient peculiar privilege of the fishers of white fish, who, by an Act of Queen Mary, were permitted to use fixed engines —a custom illegal elsewhere and on the English side of the Firth.

The nets used locally are of peculiar construction. "Paidle" nets, used for catching white fish, are fixed engines similar to salmon-, stake-, and bag-nets, but on a smaller scale. In the Annan fishery district there are 10 "white fish nets," having 24 pockets : in the Firth district 66 with 66 pockets. Halve- or haaf-nets are a kind of bag-net, 14 feet long, with three perpendicular rods under them, one at each end, and one in the middle, to keep down the net. These nets are held by men in the current of the ebbing or flowing tide. The "halvers," or fishers of Annan, claim the right to use these under a charter granted in 1538. "Poke-nets" are about a yard square at the mouth in form of an open bag, and are suspended between stakes from 6 to 7 feet long, which are fixed about half-way into the sand at a distance of 4 feet from each other. "Stake-" or "trap-nets," invented by John Little of Newbie, are nets of one or more compartments, enclosed with netting, supported by stakes, from 5 to 15 feet high or upwards, driven into the sand or beach, and with netting for their roof. They have doors which open with the inflowing tide and are effectually closed by the returning current. Two long leaders guide the fish into these traps. The Special Commissioners of the Solway in 1881 granted to four maritime proprietors and to the Burgh of Annan the privilege of using stake-, fly-, and bag-nets to the number of 34 engines and 126 pockets. Poke-nets to the number of 600 "clouts" and 2200 pockets were allowed to three maritime proprietors and to the Burgh of Annan.

"Whammelling" is a method of fishing on the Solway introduced in 1855 by two fishermen named Woodman, whereby fish are caught in a long net, 600 to 800 yards long. One end of the net, weighted on the under side and attached to a loaded pole which remains upright, is thrown from the stern of a boat moving across the tide and paying out all the net till it extends across the channel. The upper part of the net floats free with the tide, and in the meshes of the net the fish are caught. About 50 boats leave Seafield, Annan, and other places to "whammle" for salmon, "draw" for trout, "trawl" for shrimps, and "beam-trawl" for flat fish.

Another method of fishing practised at Lochmaben on the Annan is called "cross-line-fishing." The line is stretched between two fishers, one on each bank; from the line drop several smaller baited lines called "eeks."

In 1840, in the Dumfries and Stranraer district, 84 boats and 430 fishers were employed; and 1665 barrels of herring were cured. In 1901 only 169 men and 1 woman were registered deep-sea fishers. The Fishery Return (1910) shows that at present there exists at Carsethorn 1 boat under 30 feet and over 18 feet, employing 18 fishermen and boys; at Caerlaverock 20 fishermen and boys; at Powfoot 13; at Annan 6 boats over 18 feet and 10 tons, and 1 under 18 feet of 1 ton. No new boat was constructed in the district in 1909.

The quantity of fish, especially flounders, taken between Kirkcudbright and Powfoot, was 1084 cwts., valued at £604; and mussels £187. At Annan 2108 cwts., valued at £2717, were taken; and £2704 were obtained for shell-fish, being an increase of £1000 on this fishing. There was a decrease of shrimps, however. The weight of fish borne by rail was from Annan 265 tons, Dumfries 11, Dornock 6, Cummertrees 3—in all 285 tons. Forty tons left Kirkcudbright. Half the income of the Annan fishermen was from shrimps : there was a short catch of lobsters. Crabs and oysters increase, and half the value of shell-fish is from oysters.

The salmon fisheries for the Solway district are valued as follows : Annan £2917, Dee (Solway) £1231, Nith £506. These sums show a decrease on former years.

For the salmon, which ascends to the extreme limits of the county, the close time for netting is from 10th September to 24th February. The close time for angling is from 13th November to 24th February. In Hoddom Waters salmon are caught with stake-nets and poke-nets. In Annan there are fixed engines, but no sweep-nets, while rod and line are in use. In the Newbie fishings the stake-net is used. In the Nith fixed engines, sweep-net, rod and line are used. Sea-trout appears in March, and grilse early in June in the Annan and the Nith. The heaviest salmon taken in Annan in 1910 was 38 lbs., and in Nith 30 lbs.

Few counties possess such facilities as Dumfriesshire for angling, and at small cost. Annan river contains salmon, trout, pike, perch, roach, chub, eel, sea-trout, grilse, and herling; Nith, salmon, sea-trout, grayling, and trout; Esk, salmon, grilse, herling, sea-trout, and trout; the Cairn, salmon, sea-trout, and yellow trout. The streams all have river-trout, some have sea-trout and salmon. Keen fishers abound, and their interests are looked after by the following associations: Dumfries and Maxwelltown Angling Association, Dumfries and Galloway Angling Promotion and Protection Club, Esk and Liddel Fisheries Association, Mid Nithsdale Angling Association, Upper Nithsdale Angling Association.

16. Shipping and Trade.

At the beginning of the nineteenth century Dumfries and Annan were important sea-ports, doing a large home and foreign trade. To-day that commerce is at a minimum as far as water-ways are concerned. In 1809 ships registered at the Port of Dumfries, that is from Gretna to Kirkcudbright, numbered inward bound 493, with tonnage 18,985, and men 802. In 1835 there were 192 ships with 11,798 tonnage and 779 men. In 1894 there was a temporary revival of import and export trade ; but in 1909 the ships recorded are six British inward with 641 tons, and three foreign of 506, together with 15 sailing-ships with 837 tons, and two steam-ships of 58 tons registered. The total outwards was nil.

Both Dumfries and Annan formerly had a large foreign trade with America and the Continent. A writer in 1811 states that the imports of Annan were coal, lime, slate, timber, herrings, salt, West India produce and English goods discharged from 200 vessels, while 40 vessels exported grain, malt, potatoes, bacon, freestone and wood. Sixteen vessels averaging 40 tons each belonged to Annan. Formerly Annan had a large import wine-trade, while ship-building and rope-making were successful industries.

Relying on this large and increasing trade, the authorities at much expense, under Act of Parliament, had the navigation of the Nith improved and three good quays provided within six miles of Dumfries, where

80 vessels belonging to the local port loaded and discharged in 1841. A steam-boat also plied regularly from Glencaple to Liverpool with passengers, bestial, and goods. Glencaple in 1840 boasted of a ship-building yard where two vessels of over 60 tons burden were built annually. But the advent of the railroad destroyed all this local business, timber, goods and cattle being transferred to the railway ; while a new wet dock at Silloth was taken advantage of by ship-masters.

17. History of the County.

The history of this region is a narrative of strife, battery and bloodshed. The marvel is that any residue of life remained in that gory arena. Historical memorials —the camps at Birrens, Birrenswark, and other places, the inscribed stones found in these camps—remain to prove the Roman conquest. On the withdrawal of the Romans a mist falls upon the doings of the warring tribes, Selgovae and Novantae north of the Solway, and the Brigantes south of it. The district again emerges into recorded fame when, in 573, Rydderch Hael, Christian King of Strathclyde, defeated the pagan Welshman, Gwenddoleu, at Arderydd on the English bank of the Esk.

At Dawstone, in 603, King Aidan was defeated by Ethelfrid. Successive religious waves passed over this region. In the fifth century Ninian and his British school planted the cross here. In the sixth century

St Mungo (Kentigern), still remembered in a parish of that name, fixed his see in Hoddom ; and a century later St Cuthbert, whose name is preserved in local place-names, confirmed the evangelisation of Cumbria, then and long afterwards a part of England, which the Scoto-

Statuette of Brigantia, found at Birrens

Irish missionaries also visited. The incidents of the racial wars between Picts, Britons, Gaels, and Norsemen are very sparsely recorded. Arthurian legends are by some associated with the Head of the " Wood of Celydon " in Holywood.

Of the intrusion of the English into Strathclyde, among the Iro-Scottish conquerors there, of the descent of Norse vikings, Iro-Northmen (Dubhgalls), and the ultimate suffusion of Anglo-Saxons, Danes and North-men among the blended British and Gaelic population, with the result in the rise of the English speaking com-munity, we have little definite information. Halfdan, the Dane, overran this territory in 875, which the Northmen made into a pleasant colony with Tinwald (*Thingvellir*) for their local capital. The English King, Athelstan, in 937 inflicted a terrible defeat upon Anlaf, King of Ireland, and a strong combination of Scots, Welsh, and Irish, at Brunanburh. Five Kings, seven Jarls, a son of Constantine, and two brothers of Athelstan bit the dust there. According to one competent authority, this victory was won near Birrenswark. The magnificent rune-inscribed High Cross at Ruthwell, if not a witness of that battle and sea-flight, may have been its piously founded memorial. In 945 King Edmund granted Cumbria to Malcolm, King of Scotland.

William the Conqueror's sequestration and displace-ment of the Saxon lords by Norman soldiers brought the families of Brus, Jardine, Comyn, Herries, Johnston, and others into contiguity with the Borders. The sons of the soil, if local traditions are trustworthy, were Rorys, Welshes, MacMaths, Morras, Karls, Crechtons, Cruthers, Graems, Griers, Ferguses, Eggers, Irvines, and others with Celtic names. In 1107 King Edgar gave Edmund's troublesome gift to the gallant Prince David, who, with the help of his Anglo-Norman associates, ruled it till his death. For him the Scottish Borderers under Prince Henry fought at Northallerton in 1138. David gave the See of Glasgow jurisdiction over the shire and over part of Galloway. When Malcolm the Maiden ceded Carlisle to Henry II, the national boundary was shifted to the Esk. The great ford was at Sulewath, that is *sul-* or *sol-vad*—the mud-ford. Hence arose the name Solway Firth, or Scotiswath or Scotwade (*vadum Scoticum*), which was also known as *Tracht Romra*, or the Shore of the Strong Tide. The land between the Esk and the Sark became debateable, and Solway (more properly Sollome or Solane) Moss was made English when the boundary was fixed at the Sark in 1552.

The thirteenth century was noted for the munificent works of Dervorgilla, widow of John Baliol, who founded a monastery for Grey Friars in Dumfries. In the struggle with England, 1286-1371, the Dumfriesians had a share, their leaders oscillating in their allegiance between the Kings, and the people bleeding for both sides. Traditions of Wallace are still vividly narrated in folk-story, which tells how he took the castles of Enoch and Tibbers, and spilled Moreland's blood at "The Sax Corses" in Kirkmichael. So is the story of Bruce's dispatch of Comyn when Kirkpatrick did "mak siccar" his bloody work before the altar of the Grey Friars in Dumfries. King Edward I besieged and took Caerlave-rock Castle in 1300, returned to England, and died near Burgh-by-Sands as he was on his way to desolate Dumfriesshire again. Many other Kings, down to the time of King James VI, visited Dumfries bringing peace.

But the part and lot of the Borderer was war, and blood was ever in his wine cup. Border history largely turns round the names of three influential families—the Maxwells, the Douglases, and the Johnstones. Of the supersession of the Dunegal family of Nithsdale, and of the Edgars, their descendants, in favour of the March family and of the Douglases of Morton and Drumlanrig, we have charter evidence back to the fourteenth century.

When the chief of the Brus family ascended the Scottish throne, another very masterful Border chief, Maxwell, had long established himself at Caerlaverock, and his family and vassals played an important part till they were overshadowed by the Douglases, Lords of Galloway, the Knights of Drumlanrig, and the Dukes of Queensberry, and ousted from places of honour and office by Johnstones and Crichtons. About 1263 we find Sir Aymer de Maxwell Sheriff of Dumfries, and in 1409 a Maxwell becomes Steward of Annandale, and their descendants Wardens of the West Marches. There is not a little in the boast of the Johnstones :

"Within the bounds of Annandale, the gentle Johnstones ride,
    They have been there a thousand years, and a thousand years
        they'll bide."

The gallant Sir William Douglas, who married Egidia, daughter of Robert II, in 1387, with her obtained territory in Nithsdale. Whenever

"The doughty Douglas boun' him ride
Into England to make a prey,"

we find the men of Dumfries and Galloway at his

back. Archibald the Grim, son of "Good Sir James," ruled the Marches from Thrieve Castle, and was frequently on the war-path, as later (1400) his son Archibald, Warden of the Marches, was against the Earl of March and Hotspur Percy. This Douglas and Percy feud was perennial.

A few of the more striking incidents in Border warfare may be mentioned. In 1297 Sir Robert Clifford slew over 300 Annandale men at Battlefield on Annan Moor, and on a second raid in 1298 burnt Annan. In 1332 Edward Baliol was nearly captured in a fight at Annan. In March, 1333, Sir Antony Lucy defeated and captured the Knight of Liddesdale at the Battle of Dornock. The Scots had their revenge when, on 23rd October, 1448, Douglas, Earl of Ormond, with 4000 Scots met Percy with 6000 English at Clochmabenstane, and in this battle of the Sark routed the English, captured Percy, and slew 2000 of the foe with a loss of 600 men. Dumfries was burned in 1415, and again in 1449.

Between the Crown and the Douglases relations had long been strained, and at last were broken off. The King marched against the Earl, who fled to England, leaving his fiery brothers to fight it out and meet defeat at the hands of another Douglas, Angus, at Arkinholm, Langholm, on May Day, 1455. Nearly thirty years afterwards the renegade Douglas and Albany, with 500 horse, made a raid into Scotland, and in a skirmish with the Maxwells, Crichtons, Charterises, and other Border families, were captured at Kirtlebank in July, 1484. Merkland Cross marks the site where the Master of

Maxwell fell after this fight, called the battle of Kirk-connel and the battle of Lochmaben. This last Earl was sentenced to seclusion in Lindores Abbey. The Maxwells of Caerlaverock were advanced and became Wardens of the Marches and Stewards of Annandale.

During the wars with England in the sixteenth century the shire was frequently devastated. Many Dumfriesians fell at Flodden. Lord Dacre thereafter made Eskdale and Annandale into a waste. Recriminations with fire and sword followed on both sides of the Border. Rival families were at feud as well. In 1529 James V marched into the county with 8000 men to curb the Border chiefs, and hanged Johnie Armstrong and his freebooters at Carlenrig. In 1542, in revenge for a foray of the English, repelled by the Johnstones and others, James V led a force of 10,000 men to the Borders. From Birrenswark Hill, it is said, he saw his army march across the frontier to meet a disgraceful defeat at Solway Moss, on 24th November, 1542. The King went home from Caerlaverock Castle to die of a broken heart. Douglas of Drumlanrig, and the Carlyles of Brydekirk, alone of the southern Scots, would not submit to English domination ; and Drumlanrig was appointed the Warden. In 1547 Wharton, the English Warden, laid low the castle, steeple, and town of Annan as "a very noisome neighbourhood to England."

The Reformation principles, preached in the south-west by Knox, were welcomed by the majority in the shire, but not by the Maxwells. Queen Mary visited the shire several times and gathered a strong party for her

cause ; but the Regent, in turn, came and coerced them
into submission to the protestant government. Various
political parties united to oppose the English forces under
Scrope and Sussex when, in 1570, they crossed the Borders
and laid waste Annan, Dumfries, and the country around
with fire and sword.

The feud between the Maxwells and the Johnstones
had a pitiful ending when Lord Maxwell, Warden of the
Marches, was defeated and killed at Dryfe Sands on
6th December, 1593, by the Laird of Johnstone and his
followers, whom he had been authorised to apprehend.
In 1608 Sir James Johnstone, now the Warden, met
Maxwell, the son of the slain Warden, at Tinwald in
a conference for the purpose of ending the feud peaceably.
It ended differently. Johnstone was shot by Maxwell,
who fled to the Continent. On his return he was
captured and executed in Edinburgh in 1613 for high
treason and for slaying the Warden of the Marches.
James VI several times visited Dumfries, and on the
occasion of his last visit, in August, 1617, presented a
shooting trophy to the Seven Incorporated Trades, called
The Siller Gun. It was last competed for in 1901.

In the troubles consequent upon the autocratic conduct
of Charles I, and the intrusion of Episcopacy, the Lowlands were much concerned. The Covenant was generally
subscribed in 1638. Caerlaverock, held for the King,
was taken by the Covenanters in 1640. On the accession
of Charles II the south-west counties were so cruelly
treated by the military that in 1666 the men of the
Glenkens rose in Dalry, marched to Dumfries, seized

Sir James Turner, the Commander, and marched away with him as a hostage to Edinburgh to appeal for justice. The royal forces, under Dalyell, met them at Rullion Green, defeated them, and captured many for execution and banishment. Colonel John Graham, of Claverhouse, was appointed depute sheriff of the county in 1679 and

Caerlaverock Castle

later a justiciary judge, and hunted the Covenanters down. Many natives of the shire were shot, executed after trial, and exiled on account of adherence to the Covenants. At the Cross of Sanquhar the Cameronians publicly disowned Charles II in 1680, and James VII in 1685.

The Union of the Parliaments was not acceptable to

all in the southern counties of Scotland, and the Articles of the Union were publicly burnt at the Cross of Dumfries by disaffected patriots. The affair of the Old Pretender in 1715 affected only the fifth Earl of Nithsdale, who was taken at Preston, and narrowly escaped execution through an escape successfully carried out by his gallant wife. "Bonnie Prince Charlie" in 1745 in his march to and from England passed through the county, and from the burgh of Dumfries exacted a large sum of money, as well as many pairs of shoes for his soldiery.

The next important events are the advent to Ellisland in 1788 of Robert Burns, and of his subsequent residence and death in Dumfries in 1796. In 1832 cholera ravaged the shire, and 420 persons succumbed in Dumfries. In 1843 the Church of Scotland was dismembered ; and nine of the parish ministers and three ministers of unendowed churches, with a large following, threw in their lot with the Free Church.

18. Antiquities — Prehistoric, Roman, Celtic, Anglo-Saxon.

Many relics found in Dumfriesshire illustrate the various stages of development through which the inhabitants passed since

> "Wild in wood the noble savage ran."

The nature of the rocks did not afford many natural shelters for primitive man, and consequently the "finds" which tell of the pagans who used wooden and bone implements, then rough and polished stone tools, and last of all metallic weapons, instruments, and ornaments, are found in fields, mosses, and cairns. The constant turning over of the soil has made them less numerous than in other districts. Many sites of primitive villages, earth-forts, camps, crannogs, burial cists, and other remains still unexplored may prove fruitful to the antiquarian. Canoes, dug out of single trees, have been found in Closeburn Castle Loch, Lochar Moss, and Friars Carse Loch; and in 1911 two small canoes were found at Lochmaben. A group of houses, called "weems," excavated by primitive folk, is seen above The Deil's Dyke, on Townhead farm in Closeburn; and in them burnt hearth-stones, calcined iron, and a whorl were found.

Nearly every prominent eminence in the county has traces of defensive works, some having concentric rings of ditches, as at Tynron Dun (946 feet). Within the shire there are remains of 249 such forts, more than one half being in Annandale. Of these 14 are rectangular, eight probably rectangular, and 206 curvilinear; and 21 regular motes are preserved. Remains of a vitrified fort were found near Pinzarie, Tynron. On account of the scarcity of lakes, crannogs are few; but those on two islands in Loch Urr, although of a later type, show two defensive submerged gangways. In Lochmaben Castle Loch submerged structures exist, and in Loch Skene a small islet has the appearance of a crannog.

Cairns are very numerous. Those on the farm of Auchencairn, Closeburn, of immense size, and associated with the names of Wallace and Bruce, are of early construction. A congeries of cairns, over 100 in number, remains at Girharrow, Glencairn. Of stone circles, the most perfect are the Twelve Apostles of Holywood and the Girdlestanes of Eskdalemuir. The last of another circle is the historic Clochmabenstane, in Gretna, already referred to. Two rocking-stones—off the balance—are seen at Belstane, Drumlanrig, and at Glenwhargen.

The Romans left permanent monuments of their occupation in the Roman Road and in the Camps at Birrenswark and Birrens in Annandale, as well as at Overbie and in smaller outposts, of which a splendid example is found on the Waalpath, Durisdeer. The numerous inscribed stones found in Birrens threw a great light on the Roman occupation and recorded that the Second Tungrian Cohort, the First German Cohort, "called the Nervana," and part of the Sixth Legion, were stationed here in the second century. Portions of the Roman Road—the *magna via*—which entered Scotland at Gretna, made for Birrens and Birrenswark, and passed up Annandale into Lanarkshire, are still visible. A secondary branch turned off west at Gallaberry and traversed Nithsdale, emerging by the Waalpath (*wald* or wood-path) into Crawford. A side-road passed into the Cairn Valley. Another branch went through Eskdale on to Trimontium—the Eildons. One of the most inexplicable objects of antiquity is The Deil's Dyke, a deep ditch with the earth thrown up to form a breastwork fronting south and west. In places it is faced with stone. It was traced by Joseph Train from Lochryan through

Galloway into Ayrshire. Portions of it are visible at Cairn Hill, in Sanquhar, west of Mennock, in Dalveen,

Altar of Minerva found at Birrens

and on the hills east of Morton Castle. On Bellybucht Hill, Morton, and on Townhead farm, a fine stretch like a raised beach is seen. Through Annandale it is traced on its way to Nith or the Solway. The theory may be hazarded that this is a rude imitation of the Roman Wall, erected by the folk in Strathclyde, after the withdrawal of the Roman soldiery, in order to keep the Galwegian Picts in check.

Of motes and mote-sites one of the largest and most interesting is that of Jarbruck, known as The Bow-butts of Ingleston, in Glencairn. It is an oblong raised area, 200 feet long and from 40 to 60 feet broad, and is surmounted at each end by a tower of forced earth, respectively 30 and 44 feet high. It has the characteristics of a site of a Norman palisaded stronghold, or peel. Among the many watch-hills, watch-fells, watchman's-knowes, and bale-hills in the shire, the following are beacon-heights noted in the fifteenth century: Trailtrow (or Repentance Tower), Wardlaw, Trohoughton, Barlouth, Pantath, Whytewoollen, Dowlarg, Kinnelknock, The Bleize, Gallowhill, Watchfell (Closeburn), Cruffel (Sanquhar), Corsincone.

Of the restricted culture of the aborginal races we have evidence in tools from the water-worn stone up to the most carefully designed and highly polished stone-axe of greenstone. Flint arrow-tips and javelin heads, sandstone hammers, massive pierced grey sandstone hammers, axes of greenstone, compact sandstone, granite, claystone, and whin are numerous. A man digging peats in Solway Moss discovered a stone axe inserted in its handle, just as elsewhere flint tips have been found in the original arrow shafts. Adder stones, beads, whorls, and other objects in great variety, belonging to a primitive age, have been found. Many querns, barley-stones or "knockers," and pounders still lie near old habitations.

The Bronze Age is also well illustrated by the many weapons and ornaments found in this area. Socketed spearheads, daggers, sword blades of all sizes, socketed axe-heads, and other articles in bronze, in excellent condition are preserved in museums. Bronze paterae of a Roman type were unearthed near Friars Carse in 1790. A magnificent bronze pot found in East Morton was personally carried off to Abbotsford by Sir Walter Scott. A neat bronze tripod ewer was found in Glencairn, another in Keir, and another near Bonshaw Tower. A lovely example of bronze metal and enamel work is a bridle-bit found near Birrenswark. A rare specimen of a beaded collar, or neck-ring, in bronze, $6\frac{1}{2}$ inches in diameter, of late Celtic type, beautiful in design and finish, was found in Lochar Moss, lying within a gracefully shaped bronze bowl. A fillet of thin bronze, ornamented in delicate *repoussé* work, and five bosses of bronze were found in the shire. A golden collar was found in Middle Nithsdale.

Evidences of Celtic and later culture are preserved in the few fragments of cross-shafts still remaining. The pedestal of the baptismal font in Kirkconnel Church, Nithsdale, is part of a Celtic cross-shaft. Portions of the sculptured stone crosses of Closeburn, Durisdeer, Glencairn, and Penpont are preserved in the Dr Grierson Museum, Thornhill. A fragment of the cross of Hoddom, displaying a saint with a nimbus, is preserved. One of the most important relics of antiquity is the Ruthwell Cross, dashed into pieces in the seventeenth century, but now, since 1887, safely guarded within the parish church of Ruthwell under the Ancient Monuments Preservation Act. It is one of the few sculptured high-crosses remaining in Scotland. It was erected for a devotional purpose. It is a

Torque and Bowl found at Lochar Moss

free standing cross, 17 feet high. The shaft measures 10 feet 6 inches high and 3 feet 1 inch across the arms. The shaft tapers from 21 inches to 13 inches in breadth and from 18 inches to 9 inches in thickness. The stone

Ruthwell Cross

is sculptured in relief in ten panels on each side; and the sculpture represents incidents recorded in the Old and New Testaments. Inscriptions in incised Saxon capitals quote texts in the Vulgate. An Anglian runic inscription on the sides of the cross is taken from *The Dream of the Rood*, which has been attributed to Cynewulf, probably of Northumbrian origin and writing in the eighth century. Other authorities have attributed *The Dream* to Caedmon, and Stephens deciphered part of the Ruthwell runes as "Kadmon mæfaucetho," which he translated "Caedmon made me." But, two objections have been urged to this —first, that these words are mere jargon, belonging to no known or possible Old English dialect; second, these words cannot come from the runes visible on the cross. This remarkable object is reminiscent of the return wave of Celtic christianity out of England.

At a ford across the Nith at Thornhill, stands a beautiful floriated sandstone cross of fifteenth century work, in a good state of preservation, and only wanting parts of the arms. It was probably a terminal or votive cross and thus escaped the iconoclasts of the seventeenth century. The shaft measures 9 feet 2 inches long, tapers from 18 to 15 inches in breadth, and from 7 to 8 inches in thickness. It is a panelled cross with zoomorphic and dragonesque designs.

A later cross is that still standing in the baronial market-place of Moniaive, and placed there by Fergusson of Craigdarroch in 1638. On it the jougs, still preserved, were formerly affixed. A mere fragment of the historic Cross of Sanquhar is preserved in a United Free Church in Sanquhar. The Cross of Thornhill, comparatively modern, is admitted to be the handsomest market-cross in the country, and was erected about 1714 by the Duke of Queensberry. Its high, massive, fluted column, erected upon a massive octagonal base, approached by a series of

Boatford Cross and Nith Bridge

steps, and all cut out of the local red sandstone, is sur-mounted by a bronze flying Pegasus.

In Kirkpatrick-Fleming parish are preserved two crosses—one at Kirkconnel, besides the ancient church, formed from one grey stone in the form of a Latin cross, and standing 7 feet 4 inches high ; and another at

Merkland. The latter stands about 11 feet and a half high, the shaft being 9 feet high and surmounted by a pierced cross formed by the union of four *fleur-de-lis*. It is said to mark the spot where the Master of Maxwell fell after the defeat of Albany, in the neighbourhood, in 1484.

Thornhill Cross

The tombstones, once sculptured, of "Fair Helen" and Adam Fleming are to be seen in Kirkconnel churchyard, beside the Kirtle.

Roman coins of the age of Nero, Vespasian, and Domitian were found at Broomholm in Langholm.

Roman coins were got in the excavation of Birrens camp. Hoards of coins, especially of the mints of the Edwards of England, have been found in Durisdeer, Closeburn, and in other parishes. Few of the Borderers' blades which fought for Crown and Covenant escaped the search made for them under the disarming statutes of the later Stewart kings.

19. Architecture—(a) Ecclesiastical.

Dumfriesshire is unique in this respect that there is not preserved within it a single example of a Celtic, British, Saxon, Norman or Medieval church, or ecclesiastical edifice. This is one unhappy result of warfare on the Borders. The much admired abbeys of Lincluden and Sweetheart are just over the boundaries: no similar ecclesiastical edifices—churches or religious houses—of such beauty and distinction existed in the county, as far as is known. There remain well-defined sites and foundations of primitive churches indicating their smallness. The Grey Friars monastery, Dumfries, was totally razed and its stones utilised in local edifices and in St Michael's church. Similarly the abbey of the White Friars at Holywood (*Sacrum Nemus*, Dercongal), built in 1141, as well as a later Hospital, has disappeared save a few fragments. A precious bell with its Latin inscription bearing that "John Welsh, Abbot of Holywood, caused me to be made in 1505," still hangs in a modern belfry of the parish church. Of Lochmaben church, an ancient bell said to have been the gift of King Robert Bruce, is the only relic. Its Latin inscription, translated, runs "John Adam made me. Hail Mary!" The vesper bell of Dumfries, presented to the town in 1443 by Lord Carlyle of Torthorwald, is now a relic in the Observatory Museum, Maxwelltown.

A fragment of the Priory of the Canons regular of St Augustine—founded in Canonbie by Turgot de Rosse-dal, in the reign of David I—remains in the churchyard there. It is a beautiful Gothic arch, being part of the sedilia, of thirteenth century date, and forms a framework for the monument of a former parish minister. The English army destroyed the church and priory in 1542. Of the early church of St Cuthbert, at Moffat, only a part of a Gothic window with one mullion, of uncertain date, is left. Nothing visible of the ancient church of Sanquhar remains. The present edifice covers the foundations of the older structure, which excavations in 1895 proved to have been a nave and choir, 96 feet long and 30 feet 6 inches broad. The late Marquess of Bute restored to the church the effigy of a medieval ecclesiastic, probably a Crichton, rector of Sanquhar, which was long preserved at Friars Carse. An effigy, said to be that of Simon de Carruthers, is seen at Mouswald: a much later effigy lies in Morton churchyard; and three sculptured monuments of early date are shown in Dornock churchyard. At Kirkbride, Durisdeer, the shattered walls of a pre-Reformation church exist, and an early pointed window constructed of two stones. Among the *débris* lie the fragments of an effigy of an ecclesiastic who bore the

Queensberry Monument, Durisdeer Church

name of "Gabrialdus." The present parish church of
Durisdeer has a distinctive tower, church, school, and
ducal waiting-rooms, all formed out of the dark red ashlar
masonry of the obliterated castle of Durisdeer, famous in
the Wars of Independence. A similar fate befell the
"dun" of Tynron, now transferred to Tynron parish
church. The church of Durisdeer was built in the end
of the seventeenth century by the artificers who completed Drumlanrig Castle. In a mausoleum annexed to
the church an elaborate marble monument is erected to
commemorate the Duke and Duchess of Queensberry,
who died respectively in 1711 and 1709. It is reckoned
a masterpiece of Roubiliac, and the exquisite carving,
especially of lace, upon the white Carrara marble is much
admired.

The churches, built in the eighteenth century, when
there was a revival of ecclesiastical activity, are of a most
uninteresting domestic type, with the exception of St
Michael's church, Dumfries. It stands upon the site of
a very early church. The steeple was built in 1744. A
graveyard, in which Robert Burns and many distinguished
Dumfriesians are buried, surrounds the church. In the
nineteenth century many beautiful churches have been
built both in the larger towns and villages, the most
beautiful of all being the Gothic Memorial Church
erected in 1889-1898 near Dumfries, in memory of
Dr Crichton and of Mrs Crichton, founders of The
Crichton Royal Institution for the weak-minded. A
striking feature in the landscape of Moffat are the parish
church and manse, built of red Corncockle sandstone in

Early English Gothic style, and finished with great chasteness in 1887. Another handsome edifice in Moffat is the United Free Church in French Gothic. Among other fine parish churches are Morton, Penpont, Closeburn, and Greyfriars, Dumfries.

Crichton Memorial Church, Dumfries

20. Architecture—(b) Castellated.

The deficiency of Dumfriesshire in old ecclesiastical edifices is made up for by the number of its historic castles and towers, in ruins or in habitable use. Of noble piles upreared to dominate the rural scene are

Drumlanrig, "The House of the Hassock," with im-
perious front watching Middle Nithsdale ; Caerlaverock,
even in ruins still mightily menacing the Solway and the
Strath of Nith; and royal Lochmaben, holding guard of
Annandale.   No structural remains belong to Saxon or
Anglo-Norman times, except fragments encased in later
buildings.   Of fortresses in the thirteenth century only
three of first importance are mentioned—Morton, Dalswinton, and Lochmaben, and the original castles exist no
more.   An almost perfect mote, overlooking the Annan,
is all that remains to tell what a strength Annan once
was.

The oft-repaired castle of Caerlaverock, with its defended approaches, marsh, ditches, deep moat, high walls,
lofty and massive towers, is an example of a strong feudal
fortress.   It is not Anglo-Norman.   Its outer walls are
of date not later than the thirteenth century.   It is an
aggregation of works of six different periods, but in one
part or another it is an historic eye-witness of the events
of six centuries and a half.   On a mound, surrounded
with water in a moat 70 feet wide, this castle is built on
a triangular plan, the largest base being 171 feet and the
other two 150 feet in length.   Between two towers, which
are 26 feet in inside diameter and 40 feet high, built at
the apex of the triangle are situated the great entrance
gate, door, and drawbridge.   A tower stands at each of
the other points of the triangle.   A castle on a similar
plan fell to the attack of Edward I in 1300, and the present castle, while bearing marks of reconstruction shortly
after that event, has many more additions dating from the

fifteenth and sixteenth centuries. The castle also stood sieges in 1312 and in 1640 (see p. 94). It was the seat of the Maxwells, knights, lords, earls, sheriffs, stewards, and wardens of the Marches, and is now the property of their descendant, the Duchess of Norfolk.

The first castle of Lochmaben stood on the site now known as the Castlehill. The second, and existing stronghold, is built on a peninsula covering sixteen acres running into the Castle loch, and defended by ditches and a deep ashlar lined moat. Access to it was got by boat rowed into a defended ditch. In some parts it dates from before 1300. This important place stood many sieges. Edward I took it in 1298; Bruce fled to it in 1306; De Boune held it in 1346; Douglas took it in 1384; and James VI stormed it in 1588. On this last occasion the office of constable was transferred from Lord Maxwell to Johnstone of Annandale. It was granted to John Murray in 1612, and is now held by the Earl of Mansfield. In 1503–4 James IV repaired the castle and built the great hall.

Of simple vaulted towers, dating from the fourteenth century, there are two good examples, in Closeburn and Torthorwald. Closeburn Castle, formerly the seat of the Kirkpatricks, is a massive rectangular tower, founded on a mound in a lake now drained, and consisting of three vaulted stories. It measures 45 feet by 34 feet 6 inches, and rises 50 feet to the parapet. The old curiously wrought iron "yett" (gate) still hangs opposite the old entrance high in the wall on the first floor. The castle is still inhabited.

Torthorwald Castle, the ancient seat of the Carlyles,

now in ruins, also stood on a mound in a marsh, defended with a ditch.   It is also an oblong vaulted hold, 56 feet 6 inches by 39 feet 2 inches, and rising 45 feet to the apex of the highest vault.

Of keeps to which domiciliary additions were added in the period between 1400 and 1542, mention may be

Morton Castle

made of three—Morton, Sanquhar, and Comlongon. Morton, three miles north of Thornhill, stands on the site of an older castle, on a steep eminence overlooking a natural loch.   Its plan is remarkable.   Between two lofty towers access was got to an irregular oblong build-ing, whose high ashlar walls extended 92 feet till they ended in another tower at the south-east angle.   Parallel to this building was another of similar character and size. The great hall measured 93 feet by 31 feet. This massive pile had an imposing appearance. The Celtic overlord Dunegal and his powerful descendants had a castle here. From them it passed through Randolph, to the crown and to the Douglases of Nithsdale. From Sir William Douglas of Coshogle and Morton, in 1619, it passed to the Dukes of Queensberry and from them to the Scotts of Buccleuch, who hold it still.

Sanquhar keep still stands in the corner of a fort defending a courtyard into which entry is obtained from another defended yard. The domiciliary buildings attached to the keep occupy an eminence overlooking the Nith— the area of the site being 167 feet by 128 feet. The keep is a vaulted edifice measuring 23 feet square inside and with walls 10 feet thick. The Dunegal family had an interest in the place, and after them Rosses and Crichtons became the barons. In 1296 William le Tailleur was the "Warden of the new place of Senewar." In the fifteenth century a Crichton built the keep and his descendants enlarged it into a fortified residence. The Earl of Dumfries disposed of the barony of Sanquhar to the Earl of Queensberry in 1639; and his descendant, the Duke of Buccleuch, sold the castle to the late Marquess of Bute in 1896.

A very fine example of a fifteenth century vaulted house is the oblong keep of Comlongon, near Ruthwell. It measures 48 feet 10 inches by 42 feet 7 inches and rises 49 feet to the top of the battlements. The gabled rooms above the battlements and the turret afford a picturesque feature. The iron "yett" still bars the entrance. The basement vault is 17 feet 5 inches square; the hall measures 29 feet 4 inches by 21 feet 2 inches, and rises 14 feet 6 inches to the roof. With other domiciliary additions Comlongon Castle is now a residence of the Earl of Mansfield. Cockpool, or Cokepule, Tower, an older residence of the Murrays, stood a short way off.

Amisfield Tower is one of the most picturesque of the later strong houses with vaulted apartments. It is an oblong of 31 feet 6 inches by 29 feet, and has a great hall 21 feet by 15 feet. It was the seat of the Charterises, and bears the arms and initials, "I.C. 1600," of John Charteris. Built in the same period are the two ruined towers of the Johnstones, Lochwood, and Lochhouse, near Beattock. It was the inaccessible sites of such towers that made the Scottish king say that only a thief in heart could have built them there. Another fortalice of the same character with earlier foundations is Spedlins Tower, the ancient seat of the Jardines. This massive vaulted structure, repaired in 1605, situated over five miles north from Lockerbie, has a striking appearance with its projecting turrets at the four corners and deserves to be restored. Elshieshields Tower, in the neighbourhood of Lochmaben, is inhabited.

Robgill, Woodhouse, and Bonshaw Towers stand near each other in Kirkpatrick-Fleming. Bonshaw Tower, the seat of the gallant Irvings, with its separate domiciliary additions, is now inhabited. It stands on the top of a high bank above the deep vale of the Kirtle, a short distance back from the cliff. It consists of four floors with one

room on each floor, to which a good wheel stair leads.
A fine hall, lit by four windows, measures 27 feet by
17 feet 8 inches and by 10 feet 3 inches to the ceiling.
On the lintel of this hoary tower is inscribed the motto
"Soli Deo Honor Et Gloria." It is one of the few local
castles possessed and occupied by a descendant of the

Bonshaw Tower

Johnie Armstrong's House—
Hollows Tower

original holder, and chief of a Border clan. It is the
seat of Colonel Irving.

Repentance Tower is a beacon tower, built about 1562
by Lord Herries, Warden of the West March, on the site
of the old chapel of Trevertrold or Trailtrow, on an
eminence in a graveyard above Hoddom Castle and

overlooking the Solway Strath.   The founder had "a greit bell and the fyir pan put on it."   Above the doorway is inscribed the word "Repentence."   Hoddom Castle, another keep built by Lord Herries and altered in the seventeenth century, when it was sold to Sir Richard Murray, with its extensive domiciliary additions, is situated in a fine demesne and is the residence of Mr E. J. Brook.

Hoddom Castle

All these fortified residences are put in the shade by the magnificent pile of Drumlanrig, in whose foundations are encased part of "The House of the Hassock," of which the "Whig's Hole," or vault, with the iron "yett," remains as a nucleus for the vast superimposed domicile. It consists of a rectangular building 146 feet by 120 feet,

built round an open courtyard measuring 77 feet by 57
feet. The basement, which is the kitchen department,
is vaulted. By a double ram's-horn staircase access is
gained to a platform, resting on an arcade, and to the
doorway and porch under a central tower with its great
hall, 52 feet by 20 feet. Within there was formerly a
magnificent gallery 145 feet long, which is now curtailed.

Stapleton Towers

At each corner of the courtyard a circular stair gives
access to the various floors and roofs. The building was
begun in 1675 by William, third Earl of Queensberry,
and finished by him when Duke in 1689.

Among other fortalices in the county the following were
of note, few being still tenanted:—Auchen, Auchincass,
Barjarg, Blacket, Bogrie, Cornal, Cowhill, Dalswinton,

Dryfesdale, Eliock, Enoch, Frenchland, Friars Carse, Glenae, Glendinning, Hollows, Holmains, Isle, Bankend (Isle), Lag, Newbie, Raffels, Redhill, Sandywell, Stapleton, Mouswald, Tibbers, Wallace's House, and Woodhouse.

Of recent mansions in the style of a fortalice may be noted Lettrick in Dunscore, the home of Major-General Tweedie.

21. Architecture — (c) Municipal and Domestic.

The towns in the south-west of Scotland are not sufficiently large and wealthy to permit of the erection of any municipal or public buildings of the first rank, and none have survived from the past. On the main street of Sarquhar stands a very picturesque tolbooth or council-house, presenting a high tower with a conical roof, with pinnacles at the corners, and with a parapet resting on an ornamental corbel table. An outside stair gives access to a platform and from it to the tower and the chambers annexed. The buildings are of late erection—after the sixteenth century.

On the High Street of Dumfries the most prominent object is the mid-steeple, town steeple, or town house, built between 1705 and 1708, of Castledyke sandstone. The tower is said to be modelled after that formerly in the old college of Glasgow. On the south front the royal arms of Scotland, a figure of St Michael treading upon the serpent, and the standard ell, are shown sculptured. The whole edifice has lately been repaired and decorated.

The county buildings, in Buccleuch Street, consist of a very imposing edifice, in the Scottish baronial style.

The town hall of Annan is a capacious building, of baronial style, surmounted by a clock tower, from which curfew is rung every night. It stands on the site of the old castle, a short distance from the river, at the western

The Mid Steeple and High Street, Dumfries

end of High Street. Before it stands the statue of Edward Irving.

Lockerbie has an imposing town hall, with a lofty clock tower finished with a small spire rising above four turrets at the angles of the parapet. It was founded in 1887 to commemorate the Jubilee of Queen Victoria.

The assembly hall is very handsome.   Contiguously placed
is the Carnegie Library, with its supplementary suite of
rooms for amusements.

The town hall of Langholm, in the market-place, was
the gift of the Duke of Buccleuch.   Beside it stands a

Annan Town Hall

(*Showing statue of Irving*)

handsome edifice built of the white sandstone of Whita
Hill, as a public library to house the books purchased with
a bequest left by Telford the engineer.   The Thomas
Hope Hospital, Langh lm, is a large building, a prominent
feature of which is a central castellated tower, erected out

of a bequest of Mr Hope, a native of Westerkirk, for the treatment of the sick and the succour of the indigent.

Lochmaben town hall at the one end of the main street and the church at the other are the two features of the royal burgh.   In a niche above the door of the town hall is a statue to the Rev. William Graham, looking

Castlemilk

down on a white sandstone statue of King Robert the Bruce, placed there mainly by the exertions of Mr Graham.

The beautiful mansions of the county, many set in charming surroundings, are a distinctive feature, and are the residences of many landlords who take a personal interest in all local affairs.   Among the many are Cowhill,

Comlongon Castle, Crawfordton, Castlemilk, Duncow, Elshieshields, Friars Carse, Jardine Hall, Langholm Lodge, Newtonairds, Raehills, and Stapleton Tower.

Friars Carse, Dunscore

22. Communications — Past and Present. Roads and Railways.

In comparison with other districts this territory, since the occupation of the Romans, has enjoyed splendid means of communication. The great western Roman road out of Caesariensis Maxima into Valentia passed through Carlisle on to Longtown, in Cumberland. Here one branch went by Netherby to Liddel Moat up Eskdale to Castle Oe'r and Raeburnfoot onward to Trimontium or The Eildons. Another branch crossed the Sark at Burghslacks near Gretna, passed into Kirkpatrick Fleming, on by Birrens, through Hoddom by Lockerbie to Gallowberry above Lochmaben. Here the road branched to the west towards Nithsdale, past Lochmaben, Tinwald, Duncow, Closeburn, Thornhill, Durisdeer village to Crawford. Smaller branches traversed Kirkmichael, Glencairn and Tynron. The main road followed the Annan by The Devil's Beef Tub to Crawford, where it joined the Nithsdale branch again. Remains of this built road are traceable and have been laid bare in many places.

After the passing of the Turnpike Act in 1777, by which rates were leviable, great impetus was given to road-making and bridge-building. Old thoroughfares, which under the Act of 1686 were badly kept up by the tenantry, were widened, drained, and fenced in, so that in the beginning of the nineteenth century Dumfriesshire was intersected by good mail and coach roads. Among those then interested in road-building and navigation were Major-General Dirom, Mr Maxwell of Springkell, Patrick Miller of Dalswinton and Telford, the engineer. A new turnpike road from Glasgow, leading down Evandale to Beattock Bridge, was continued by Lochmaben and Annan to Carlisle. Another was made through the vale of Carron to Elvanfoot. Later still another was laid from Moffat into Nithsdale, and one from Springkell to Kelhead.

The main road ran from Carlisle to Dumfries, then through Galloway to Portpatrick, so that there was every facility for the transference of goods, transport of passengers, and driving of cattle. Old drove-roads traversed hill and dale in all directions, and these—such as that still visible in Enterkin Pass, by which the soldiers of Claverhouse took the Covenanters to Edinburgh—were suitable for foot passengers, pack-horses, horses with sledges, and droves. Sir Charles Stuart-Menteith introduced stone causeways for steep gradients on his Closeburn estate, and early in the nineteenth century plans were prepared for an iron railway between Dumfries and the coalfield in Sanquhar, for a canal between Dumfries and Carlisle, and for other canals from Powfoot to Lochmaben, Dalswinton to Caerlaverock, and from Annan to Kirkbank.

There are many passable fords on the rivers and streams. But in 1812 there were 16 good bridges in the county—six over Nith, five over Annan, and five over Esk. All that can be said of the hoary bridge, which, without documentary proof or ancient tradition, is associated with the name of Dervorgilla as its supposed foundress, is, that for centuries at least this historic bridge has spanned the Nith at Dumfries, and is a memorial of the value of the ancient highway into Galloway. The lovely old bridge at Drumlanrig, ingeniously improved by Charles Howitt, half a century ago, is a structure whose history is lost in antiquity. Another substantial bridge crosses the Nith at Boatford, near Thornhill, and, as the inscription on its parapet bears, was built by "William Morton, Master Mason, 1777." So early as 1560 a bridge existed in Moniaive, and in the seventeenth century Fergusson of Craigdarroch built a bridge of one arch over the Dalwhat, and to this another arch has been added. Through Moniaive ran the route of the Craigengillan coach, which did the journey between Dumfries and Glasgow, by way of Carsphairn, in thirteen hours and three quarters. Other "roaring dillies," as they were called, plied between Carlisle and Edinburgh and Glasgow through Nithsdale and Annandale.

A new bridge at Dumfries, founded in 1791, was completed in 1794; another handsome bridge over the Annan in that burgh was opened in 1826. The bridge at Dumfries has, within recent years, been widened and beautified at the private expense of a lady citizen, and is now an elegant and broad viaduct. But all these roads and bridges are insignificant compared with the massive stone embankments, and the vast viaducts of stone and iron required by the railways for crossing the Nith, Annan, Carron and Cample, and especially the magnificent iron viaduct for the railroad across the Solway at Annan.

Before the introduction of steam-traction the Nith and the Annan, with the Solway estuary, formed a natural waterway of first commercial importance. Large sums of money have been expended on making these waters navigable. There are no canals and no tramways in the county.

By the introduction of the locomotive the system of transport was revolutionised, and "droving" was superseded. The Caledonian Railway from Glasgow and

Old and New Bridges, Dumfries

Edinburgh to Carlisle was opened for traffic in 1847. Its route is through rocky Evandale and down Annandale, and it serves all the vales converging on Moffat at Beattock, Lockerbie, Ecclefechan, Annan, and Gretna Green. A short branch line connects Beattock to Moffat. At Annan the line crosses the Solway into England. A keen parliamentary fight occurred over the question of route through Annandale or Nithsdale, one line at first being considered ample from a paying standpoint.

This undertaking was soon followed by the Glasgow and South-Western Railway, which was completed in 1850. Its tract is through Ayrshire into Dumfriesshire at Kirkconnel, through Sanquhar and the grand pass below it, through Carronbridge tunnel—a thousand yards long, and an early triumph of engineering—past the lovely woods of Drumlanrig, down Nithsdale, past Dumfries and Annan into Carlisle station, where the Caledonian and North British Railways also meet. The next district tapped was Kirkcudbright, by a branch from Dumfries to Kirkcudbright, in 1859—later extended to Stranraer and Portpatrick. Another useful branch on the Caledonian system is that connecting Dumfries and Lockerbie and serving the districts round Locharbriggs, Amisfield, Shieldhill, Lochmaben, and Lockerbie stations.

The North British system from Edinburgh to Carlisle skirts the county on the south-east frontier, where, at Riddings junction, a branch line runs north by Canonbie and Gilnockie (for Claygate) to Langholm. Further south from the main line at Longtown, a short branch leads to Gretna.

A light railway from Elvanfoot to Leadhills, built in 1901, has been extended to Wanlockhead. Another light railway—opened in 1905—part of the Glasgow and South-Western system, runs up the vales of Cluden and Cairn through Dunscore to Moniaive, and opens up these beautiful pastoral vales.

23. Administration and Divisions— Ancient and Modern.

The distinction which some Border land-owners claim of still being heads of ancient families, on the tribal land in their possession, is a relic of the primitive epoch when chiefs ruled tribes and were a law unto themselves, and a kind of "Jeddart Justice" prevailed. There were many baronies in this shire in the olden time, and, indeed, the present Queensberry Estate is composed of parts of 14 which have been purchased. In these chieftainries, or baronies, during the troublous days before 1603, justice was at times maintained with little deference to the feudal law that a representative of the Crown should sit in courts awarding the penalty of death. A clan system survived into the fifteenth century, for we find David II appointing a captain to the clan MacGowin, and Robert III confirming another officer in the office of Toshachdarrach— Coroner of some part of Nithsdale. The separate "law of Galloway" was another survival which held on to the fourteenth century. William the Lyon's "Judges of Galloway" sat in Dumfries.

As already stated, Cumbria had princely government for 17 years, together with local government by Celtic chiefs. Feudal government by steward, vice-comes or sheriff, and bailie succeeded; and the royal justiciar perambulated on circuit. In 1189 Udard of Hoddom acted as steward of Annandale for the Bruce; in 1264–6 Eymeric de Maccuswell acted as "Vice-comes de Dumfreis" and collected the fiscal revenues of Dumfries and Galloway. When the Celtic *maormors*, or hereditary provincial rulers, disappeared can only be surmised. From the time of David I onward the feudal system of administration, generally speaking, prevailed. In Edward I's occupation of southern Scotland his vice-comes had jurisdiction both in Dumfriesshire, as far as Eskdale, and in Kirkcudbright. Under the Scottish kings, as shown in Chapter 1, there were three officers of the Crown having jurisdiction in the shire of Dumfries, and this arrangement continued till 1748.

Burghal law was dispensed in Dumfries from the time of William the Lyon; in Lochmaben and Annan from the time of King Robert the Bruce; and in Sanquhar, the other royal burgh, from 1598, when James VI gave the barony its charter. Burghal law met all cases save those reserved in the four pleas of the Crown. The magistrates had jurisdiction over all except the King's servants in his castle, for whom a separate court was erected.

There was also a written code of laws applicable to residents on the Borders, drawn up in 1248–9, and repeatedly revised by commissioners from both kingdoms.

Wardens of the Marches—east, middle and west—were appointed for both kingdoms, and these met, down to the union of the Crowns, to see that these Laws of the Marches were duly carried out.

At present (1911) the Justices of the Peace, who have interests in the shire, from the Lord High Chancellor downwards, are numerous. The Duke of Buccleuch, Lord Lieutenant, and Keeper of the Rolls of the Peace, is assisted by three deputy lieutenants (two are baronets), by one marquis, three earls, one baron, one lord, one master, five baronets, two knights-bachelor and over two hundred justices, some of whom are privy-councillors, and many holders of land in the county. Under the Territorial and Reserve Forces Act, 1907, the Lord Lieutenant is, *ex-officio*, president of the Territorial Association. The total strength of the 5th K. O. S. B. (Dumfries and Galloway Territorials) is about 900, and of the eight companies, four are in Dumfriesshire. The 3rd King's Own Scottish Borderers (special reserve), 520 strong, also have their headquarters in Dumfries.

Under the Local Government (Scotland) Act of 1889 a County Council, consisting of 38 district members and four members representing three burghs, transact their proper business. They meet in five districts—Dumfries, Thornhill, Annan, Lockerbie, and Langholm—where representatives of 40 parish councils and of four burghs assemble when the business concerning roads and bridges is dealt with. Under the Licensing Acts, 43 representatives sit in the same districts, and Appeal Courts are held in Dumfries and Annan.

The management of the poor and of certain cemeteries
and churchyards is in the hands of 42 parish councils.
These councils superseded the parochial boards which were
instituted in 1845.   The commissioners of supply, whose
duties are somewhat curtailed since the creation of county

Old Grammar School, Annan

councils, meet to appoint their representatives on the
standing joint committee, to which the police management
is entrusted.   A joint committee of the county council
and of the burghs administer the acts relative to weights,
explosives, foods, and drugs.

Among other administrative boards are the County Road Board, the District Lunacy Board, the District Fishery Board, Nith Navigation Commissioners, Dumfries and Maxwelltown Waterwork Commissioners, Annan Harbour Trust. There is a schoolboard in every parish. Dumfries has a burgh schoolboard and a landward school-board. There are academies for higher education in Dumfries, Annan, Langholm, Lockerbie, Moffat, and Wallacehall (Closeburn).

Dumfriesshire sends one member to parliament. Its four burghs combined with Kirkcudbright also send one.

The law is dispensed by a sheriff-principal, who is now an appellate judge, by a sheriff-substitute, and by several honorary sheriff-substitutes. The sheriff holds court in Dumfries and also in Annan. The constabulary consists of 45 men and seven watchers employed for fishery protection.

The burgh of Dumfries is divided into eight wards, three councillors sitting for each of seven wards, and four for the eighth ward. There is a Burgh Licensing Court and an Appeal Court. The Town Councils of Annan and Lockerbie consist each of a provost and 11 councillors. Sanquhar, Lochmaben, and Moffat have each a provost and eight councillors.

In early Christian times Dumfriesshire formed a part of the extensive see of St Mungo (Kentigern), which extended far into England—to the Rere Cross of Stane-more. In medieval days it remained in the see of Glasgow; and its affairs were administered in the deaneries Nithsdale, Annandale, and, later, of Eskdale—Nithsdale with 21 churches, nine being beyond the shire, and Annandale with 10 churches. Some of these churches were attached to houses in Lesmahagow, Holyrood, Fail, Holywood, Kilwinning, Kelso, Melrose, Jedburgh, and

Town Hall, Lockerbie

Lincluden. Others were prebends and mensal churches of Glasgow.

At the Reformation the churches were referred to as those of Nithsdale and Annandale. The Synod of

Dumfries is mentioned in 1591; the Presbytery of Dumfries in 1593. The Presbytery of Middlebie in 1638 is turned into the Presbytery of Annan in 1742. Sanquhar Presbytery, mentioned in 1607, becomes Penpont in 1638, in which year also Lochmaben appears.

Ecclesiastical business is at present transacted in the Kirk Sessions of 43 civil parishes and five *quoad sacra* parishes, within the five presbyteries of Lochmaben, Langholm, Annan, Dumfries, and Penpont. The Appeal Court is the Synod of Dumfries, which meets in Dumfries.

24. The Roll of Honour.

The distinguished men of Dumfriesshire have been numerous, and the bede-roll of fame is large. The Kindly Tenants of The Four Towns of Lochmaben have pride in their tradition that King Robert the Bruce was born in their castled town, which he elevated into a royal burgh. Of the innumerable leaders of patriots who fought and fell for national liberty between the Lowthers and the walls of York mention need only be made of the Maxwells, Douglases, Kirkpatricks, Crichtons, Johnstones, Jardines, Carlyles, Graemes, Irvings, Elliots, and Armstrongs, and many a gallant deed will be recalled. The Western March had in the Maxwells the doughtiest of Wardens; of these none is better deserving of remembrance than the fourth baron, Lord Maxwell, who in 1543 secured from parliament the right for the lieges to peruse the Bible in their vernacular tongue. Many of the Douglases of Drumlanrig attained to great eminence as soldiers and statesmen, and among these James, second Duke of Queensberry (1662–1711), played a most important part in that critical period which ended in the union of the parliaments. Bonshaw Tower is owned by an Irving, chief of a redoubtable Border clan, from whom sprang many famous men, including Edward Irving and Washington Irving. Hector Boëthius, Scottish historian and first Principal of King's College, Aberdeen, was of Annandale descent, being related to the family of Boys— "Baro de Dryfesdale." "Rare Ben Jonson" was of the stock of Annandale Johnstones. Robert Jonston, author of the *Historia rerum Britannicarum*, published in Amsterdam in 1655, hailed from Annandale.

Ballad literature has invested with a kind of glamour Johnie Armstrong (1529) "sum tyme callit Laird of Gilnockie," and his fearless freebooting tribe. Of nobler stamp were "the four knights of Eskdale," born at Burnfoot in Westerkirk. They were Sir Pulteney Malcolm (1768–1838), admiral, who fought under Nelson, and at St Helena guarded Napoleon, who said of him, "Ah! there is a man!"; Sir John Malcolm, G.C.B., M.P. (1769–1833), the Indian administrator and diplomatist, who no less distinguished in the East by his sword than known in the West by his pen, gave us *The Political History of India, The History of Persia*, and *The Life of Robert, Lord Clive*; Sir Charles Malcolm (1782–1851) who saw much service in the East and West Indies; and Sir George Malcolm (1818–1897), who fought in Scinde, the Sikh War, Indian Mutiny, and Abyssinia. Born also in Eskdale were

Admiral Thomas Pasley (1734–1804) who assisted Howe to defeat the French fleet in 1794, and Sir Charles Pasley,

Sir John Malcolm, G.C.B.

R.E. Archibald Johnston, Lord Wariston (1611–1663), one of the Covenanters who opposed Charles I and helped

Alexander Henderson to frame the National Covenant, was born in Edinburgh, but his father was James Johnston of Beerholm, Kirkpatrick Juxta. To another Lord-Advocate and Lord of Session, George Young, a native of Dumfries, Scotland is indebted for the Education Act of 1872.

One of the remarkable results of the cessation from the tumult of Border war was the devotion with which the once bellicose inhabitants turned to peaceful avocations and to letters. They "beat their swords into ploughshares, and their spears into pruning hooks," so effectually that it is not easy to obtain a genuine blade which has drawn blood. Innumerable scholars have emerged from the country schools of the county. According to some John Duns Scotus took his oath and habit of St Francis in Dumfries; and John de Sacro Bosco, a scientific writer, hailed from Holywood. High on the long roll of literary men is "The Admirable" Crichton, James, the son of Robert Crichton of Eliock and Cluny, Lord-Advocate of Scotland and Lord of Session. He was born in 1560 in Eliock House, Sanquhar. The attainments of this prodigy—the most learned graduate of St Andrews University, master of many languages, equally proficient in every branch of learning, arms, and culture—are almost incredible for a youth who was unhappily slain at Mantua in his twenty-second year. But they have been equalled by the marvellous attainments of another Dumfriesian, Dr William Hastie (1841–1903), born in Wanlockhead, Professor of Divinity in the University of Glasgow. Gifted with an extraordinary memory, he not only

equalled Crichton as a linguist, but knew several oriental
tongues, discussed with exactitude history, theology,

Robert Flint

philosophy, jurisprudence, and science, and showed merit
as a poet. Of almost equal magnitude of powers, but in

all respects a greater thinker and author, was Dr Robert Flint (1838–1910), of Annandale extraction, a native of Dumfries, minister in Aberdeen and Kilconquhar, Professor of Moral Philosophy and Political Economy in St Andrews, and Professor of Divinity in Edinburgh. His mind was encyclopaedic. His *Philosophy of History*, *Theism*, *Anti-theistic Theories*, *Agnosticism*, and *Socialism* are likely to remain standard works.

Of other ministers of religion, whose names are worthy of honour, John Welch or Welsh (1570–1622), a reputed native of Collieston, Dunscore, son-in-law of John Knox, suffered exile in France rather than submit to the interference of James VI in Scottish Church affairs. A worthy successor of this defender of the faith was James Renwick (1662–1688), a native of Moniaive, educated in Edinburgh and Groningen, who was the last of the Hillmen judicially executed for "Christ's Crown and Covenants" in 1688. For being concerned in the Rescue at Enterkin the following Covenanters from Dumfriesshire were executed in 1684—Thomas Harkness, Andrew Clark, Samuel McEwen, and Thomas Wood. Other local martyrs, buried in the shire, were—William Smith in Tynron, John Gibson, Robert Edgar, and James Bennoch in Glencairn, William Grierson and James Kirko in Dumfries, Daniel Macmichael in Durisdeer, and Andrew Hislop in Eskdalemuir. The gravestones of the Covenanters were long kept in repair by "Old Mortality," Robert Paterson (1716–1800) who, after learning his trade, stone-cutting, in Corncockle Quarry, leased Gatelawbridge Quarry. He died at Bankend, Caerlaverock.

The most gifted preacher born in the shire was
Edward Irving (1792–1834), a native of Annan, and
educated there and in Edinburgh. He became assistant
to Dr Chalmers. He befriended Carlyle. Genius, oratory,
and a majestic figure made him the most arresting person-
ality of his age. His herculean labours disturbed his
mental equilibrium and led to the expression of doctrines
incompatible with his position in the National Church,
from whose ministry he was deposed by his co-presbyters
in Annan in 1833. Dr Andrew Mitchell Thomson
(1779–1831), born in Sanquhar Manse, became minister
of St George's, Edinburgh, where he shone as a preacher,
as in the assembly he excelled as a debater and evangelical
leader. Another doctor of the same name became minister
of Broughton United Presbyterian Church, Edinburgh,
and died there in 1901. Two other contemporary
men of mark were Dr Robert Gordon (1786–1853),
a native of Glencairn, minister of the High Church,
Edinburgh, and a writer of merit; and David Welsh
(1793–1845), born at Ericstane, Professor of Church
History in Edinburgh, one of the founders of *The North
British Review* and its first editor. Robert Johnston
(1807–1853), born in Moffat, found a mission field in
Madras. Dr John G. Paton, a native of Torthorwald,
spent a life of unwearied devotion amid perils as a
missionary in the New Hebrides till his death in 1907.

Of physicians not a few eminent men were born in
the county. Dr John Hutton became first physician to
King William and Queen Anne, and left handsome
bequests to his native parish of Caerlaverock. Dr James

Edward Irving

Mounsey and Dr John Rogerson became court doctors in St Petersburg; Sir Andrew Halliday served through the Peninsular War; and Sir William Rae, his contemporary, became Inspector-General of Hospitals and Fleets. Dr James Crichton, a native of Sanquhar, amassed a large fortune in India, where he was physician to the Governor-General; and after his decease his widow, Elizabeth Grierson, a daughter of the Laird of Lag, left the money to found The Crichton Royal Institution for Insane, in Dumfries. Dr James Currie, born in the Manse of Kirkpatrick-Fleming, gave the world the *Life of Robert Burns* in 1800. Dr William Beattie, a native of Dalton, wrote the life of Thomas Campbell, the poet. Dr John Carlyle, brother of Thomas, translated Dante. Sir William Jardine, the seventh baronet of Applegarth, distinguished himself as a naturalist and as joint-editor of the *Edinburgh Philosophical Journal*.

Westerkirk was the birthplace of that remarkable engineer, Thomas Telford (1757–1834), a genius largely self-taught, whose gifts lay in overcoming engineering difficulties and triumphing over nature, as he did in the Menai Suspension Bridge, St Catherine's Docks, Highland Roads, Caledonian Canal, Glasgow Bridge, and other monuments to his merit greater than the memorial in Westminster Abbey. Meanwhile his contemporary, Patrick Miller, Laird of Dalswinton, assisted by Symington and Hutchison from Wanlockhead, was experimenting in Dalswinton Loch with a steam-boat; and Robert Burns crossed the Nith from Ellisland to be a passenger in it on 14th October, 1788, along with Henry Brougham,

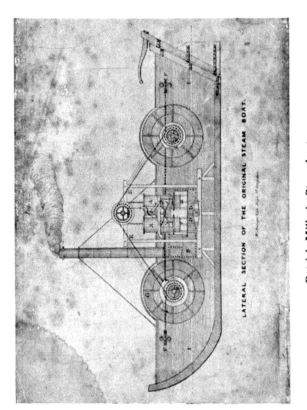

LATERAL SECTION OF THE ORIGINAL STEAM BOAT.

Patrick Miller's Steamboat

afterwards Lord Chancellor. George Graham (1822–1899), chief engineer of the Caledonian Railway, is worthy of remembrance for his enduring work in connection with that system. He wrote the *Tourist's Guide* to the system. He drove the first passenger engine from Beattock to Carlisle on 10th September, 1847. Kirkpatrick Mac-Millan (1813–1878), a native of Barflitt, Keir, invented a "dandy horse" and the first gear-driven bicycle. His smithy is at Courthill, and his burial place in the old churchyard of Keir. William Paterson (1658–1719), a native of Skipmyre, Tinwald, was the founder of the Bank of England, and promoter of the Darien Expedition. Dr Henry Duncan, parish minister of Ruthwell, in 1810 founded the Savings Bank. He is also entitled to remembrance as the protector of the Ruthwell Cross and the discoverer of reptilian footprints in Corncockle sandstone. His statue adorns the Savings Bank in Dumfries.

Many travellers have gone to arctic and tropical regions. Sir John Richardson, M.D. (1787–1865), a native of Dumfries, as surgeon and naturalist, accompanied Franklin to the polar regions, and also went in search of his lost leader. Hugh Clapperton (1788–1827), a native of Annan, was sent to Nigeria by the British Government, but succumbed in the tropics. More daring and more successful were the explorations of Joseph Thomson (1858–1895), a native of Penpont, who learned stone-hewing in Gatelawbridge Quarry, and science in Edinburgh University. At 19 years of age he went as geologist and naturalist with the Geographical Society's Expedition to Central Africa. The death of his leader left the youth in charge of the expedition, which he took and brought back without mishap.   He explored Masai

Joseph Thomson

Land, Congo Land, West Africa, and Morocco, and wrote delightful narratives of his travels.   The hardships of travel cut him off in early manhood.   His monument is erected in front of Morton School (see p. 170), where he received most of his education. Dr James Dinwiddie (1746-1815), a native of Tinwald, travelled through China. James Anderson (1824-1893), a native of Dumfries, Captain of the "Great Eastern," was knighted for his scientific services in laying the Atlantic Cable in 1866.

Of the men of letters belonging to the county Thomas Carlyle, that literary giant, naturally comes first. The son of a stone-mason, he was born in 1795 at Ecclefechan, and was in 1881 laid to rest a few yards from the Arch House, where he first saw light. In him were concentrated all the highest qualities of that unconquerable Border tribe which gave him his name, and he has requited the gift with his universal fame. Another contemporary native of Annandale, an advocate, theologian, and author of some merit, bore the same name. Dr Alexander Carlyle (1722-1805), "Jupiter Carlyle," was born in the Manse of Cummertrees. In Hoddom parish was the residence of Charles Sharpe, a writer of verses, and his better known son, Charles Kirkpatrick Sharpe (1781-1851), a quaint antiquary, a redactor of Border minstrelsy, and a friend of Sir Walter Scott.

Peter Rae (1671-1748), a native of Dumfries, minister of Kirkbride and Kirkconnel, the maker of an ingenious astronomical clock, preserved in Drumlanrig Castle, is best known as the author of a *History of the Rebellion in the year* 1715.

Among other learned men who adorned the teaching profession may be mentioned Thomas Gillespie, Closeburn,

Principal John Hunter, Closeburn, Dr A. R. Carson, Holywood, Dr Alexander Reid, Morton, Dr George

Thomas Carlyle

Ferguson, Tynron, and his brother Alexander. David Irving (1778–1860), librarian of the Faculty of Advocates and author of the *Lives of the Scottish Poets*, and *History of Scottish Poetry*, was born in Langholm. James Hannay, Dr Robert Carruthers, Thomas Aird, John MacDiarmid, and William McDowall all maintained the high traditions of literary journalism in Dumfries. Joseph Train, author and antiquary, gathered up the traditions of the south for transmission to Sir Walter Scott, who gave them immortal settings in such novels as *Guy Mannering*, *Old Mortality*, *The Bride of Lammermoor*, and *The Abbot*. One of Train's works was *The Buchanites*, a biography of Mrs Elspeth Buchan and her fanatical following, who settled at New Cample, Closeburn, as a prelude to her intended apotheosis at Templeland.

This incident suggests the observation that few Border women have been placed upon the roll of fame. Minstrels only remind us that the traveller

> " May in Kirkconnel churchyard view
> The grave of lovely Helen,"

and that on Maxwelton Braes, in Glencairn, "bonnie Annie Laurie" (1682–1764) gave "her promise true."

Too numerous for mention even are all the native poets who have sung the praises of the hills and dales of the south, down to Alexander Anderson (1848–1909), "Surfaceman," a native of Kirkconnel. The same parish is the birthplace of James Hyslop (1798–1827), teacher, and author of the beautiful poem, *The Cameronian's Dream*. The Muse in Kirkconnel also touched William Laing, as in Durisdeer it found Francis Bennoch, and in Moniaive William Bennet. Thomas Blacklock (1721–

1791), the blind poet-preacher, friend of Hume, Beattie, and Burns, was born in Annan, educated at Edinburgh University, and became for a time minister of Kirkcudbright. William Julius Mickle (1735–1788), son of the minister of Langholm, corrector of the Clarendon Press, Oxford, is best known as the chaste translator of the *Lusiad*, author of *Cumnor Hall*, and reputed author of " There's nae luck aboot the hoose." John Mayne (1759–1836), born in Dumfries, learned printing, and became a versifier. In his youth he wrote an interesting poem, with local colour, entitled, *The Siller Gun*. He settled in London, and became joint-editor and proprietor of *The Star* newspaper. A man of genius in London, contemporary with Mayne, was Allan Cunningham (1784–1842), a native of Blackwood Estate, Keir, who laid aside his stone-mason's tools for the pen of a literary man in the Capital. He enjoyed the friendship of Scott, Hogg, and Carlyle, and the patronage of Chantrey, the sculptor. Of his voluminous writings the best known are, *Remains of Nithsdale and Galloway Song*, *Lives of the most eminent British Painters*, his edition of Burns, and *The Songs of Scotland*. His own fine lyric, " A wet sheet and a flowing sea," indicates the musical genius of this able writer. Nor must another contemporary lyrist be forgotten, the Rev. Henry Scott Riddell (1798–1870), son of a shepherd in Sorbie, Langholm, who wrote the charming song *Scotland yet*.

The Corrie, the Kirtle, and the Milk had each a bard—Johnstone, Graham, and Thom—whose harmonious verse gives evidence of taste and poetic inspiration.

Burns's Monument, Dumfries

It is worthy of notice that many descendants of Borderers who gallantly bore the sword in defence of their fatherland—Douglases, Johnstones, Carlyles, Irvings and Bells—have successfully wielded the author's pen in praise of the charms of the shire. A noticeable feature of their minstrelsy is the glamour with which the streams have held these singers, as well as such poetic visitants as Fergusson, Wordsworth, Hogg, and Walter C. Smith. Every dale has had its devotee—Crawick its Laing, Wanlock its Reid, Esk its Park, Dryfe its Gardiner— happy in declaring with Bell, the minstrel of " Annan Water, sweet and fair,"

> " Steeped in boyhood's golden dream,
>   Magic lights and shadows,
>   Sing for aye, enchanted stream,
>   Through enchanted meadows."

Miller's *Poets of Dumfriesshire* is an anthology for the shire.

But all these men of letters pale their ineffectual fires before that of the King of Scottish Song, who has made Dumfries the Mecca of all true poets, since the time when Wordsworth, with his pen dipped in tears, told the educated world that Ellisland, in Dunscore, enshrines the memory of the farmer, and St Michael's churchyard, Dumfries, guards the ashes of the poet—Burns.

25. THE CHIEF TOWNS AND VILLAGES OF DUMFRIESSHIRE.

(The figures in brackets after each name give the population in 1911, and those at the end of each section are references to the pages in the text.)

Annan (4219) is a royal burgh, within a parish (6261) of the same name, on the river Annan. Its original charter, destroyed by fire, was confirmed and renewed in 1538 and 1612. Its mote, castle, and fortified steeple played a great part in Border war. Carlyle was educated at the grammar school before going to the University of Edinburgh (see p. 147). There are flourishing industries in the town and vicinity—agriculture, fisheries, grain-milling, saw-milling, wood-working, engineering, tile and brick-making, distilling, boat-building, sandstone-quarrying, and cattle sales. There is also a small shipping trade. The mansion houses of Mount Annan, Warmanbie, and Northfield are in the parish. (pp. 4, 6, 12, 15, 18, 20, 46, 47, 50, 73, 76, 77, 81, 82, 83, 84, 85, 91, 92, 93, 111, 120, 121, 125, 126, 128, 130, 131, 132, 133, 135, 140, 141, 145, 150.)

Applegarth and **Sibbaldbie** (821), is a parish in Mid-Annandale. Jardine Hall and Dinwoodie Lodge are in the parish. (p. 20.)

Bankend is a hamlet with quaint, thatched houses, in the parish of Caerlaverock. (pp. 94, 140.)

Brydekirk (951) is a *quoad sacra* parish with a village of the same name three miles from Annan. The old chapel is situated near the Annan above the village. In the red sandstone quarry of Corsehill 130 men are employed producing over 15,000 tons of stone annually. (pp. 20, 92.)

Canonbie (1838) is a parish in Eskdale. Coal is worked near Rowanburn—the Old Colliery and Blinkbonny—and 18,000 tons are banked by 105 men and boys. The limestone

Jardine Hall

quarry in the parish produces 2000 tons of lime. The Glenzier sandstone quarry is now little worked. There are five villages—Rowanburn, Bowholm, Claygate, Hollows, and Evertown—and several old towers exist in the district. (pp. 10, 21, 32, 41, 60, 69, 79, 107, 128.)

Carronbridge (200) is a village in Morton parish, whose neat homes are mostly occupied by workers on the Queensberry Estate. (pp. 15, 16, 128.)

Carrutherstown, a village in Dalton parish (589), has a public library and hall. The following mansion houses are in the neighbourhood, Dormont, Rammerscales, Whitecroft, Denbie, Kirkwood, Hetland.

Closeburn (1244) is a parish, with a village (200) of the same name. Hogg, the Ettrick shepherd, tended sheep at Mitchellslacks, formerly the home of the Harknesses, Covenanters. Wallace Hall Academy, a higher grade school, was founded by John Wallace in 1723. "The Kilns" have produced lime for nearly a century and a half. The village is known as "The Crossroads." (pp. 15, 24, 33, 35, 40, 67, 70, 72, 76, 80, 96, 99, 100, 106, 110, 112, 124, 147.)

Cummertrees (1027) is a parish, with a village of the same name. Here are situated the lime quarry of Kelhead, Kinmount House, and Repentance Tower. The fisheries give employment to some of the inhabitants. Powfoot (*q.v.*) is also in the parish. (pp. 47, 50, 80, 83, 147.)

Dornock (820) is a maritime parish in Annandale, with a village of the same name. Blacket, Robgill, and Stapleton are mansion houses in the parish. (pp. 47, 51, 83, 91.)

Dumfries (16,061), on the east side of the river Nith, as a royal burgh dates from 1186. Courts, justiciary, sheriff, burghal, and justice of peace, are held in Dumfries. The inhabitants are engaged in agriculture, floriculture, arboriculture, weaving of tweeds, hosiery and gloves, tanning, coach-building, jam-making, printing, and other industries and trades connected with these. Markets are held on the High Street, in auction-marts, and on the White Sands. Its public buildings, county buildings, town hall, academy, Ewart public library, hospitals, Crichton institution, churches, are handsome edifices. The streets are well paved, and in the suburbs, roads, wide and well laid out, lead to beautiful villas. The mid-steeple is a historic landmark on High Street, and the statue of Burns is an arresting object on the same street. In St Michael's churchyard, a befitting shrine holds the ashes of the great poet who often walked "The Plainstanes" of Dumfries. Beside his mausoleum, amid a thicket of monuments, lie three grave-slabs commemorating Welsh, Grierson, and Kirko, martyred Covenanters. In the Council Chamber is exhibited the Siller Gun, a shooting trophy presented to the Incorporated Trades of

Dunscore Church

Dumfries by James VI in 1617. The local museum at the observatory, across the river in Maxwelltown, contains many curious local relics. There are clubs for golf, bowling, hockey, cricket, carpet-bowling, lawn-tennis, rowing, football, and curling. Vessels of 300 tons discharge at Kingholm Quay. Castlebank overlooking Kingholm is the site of Dumfries Castle. (pp. 2, 3, 4, 8, 10, 12, 16, 18, 35, 36, 47, 54, 55, 58, 60, 61, 65, 66, 67,

73, 74, 76, 77, 83, 84, 85, 89, 90, 91, 93, 94, 95, 106, 107, 109, 110, 119, 120, 125, 126, 129, 130, 131, 133, 135, 138, 140, 145, 147, 149, 150, 151, 152.)

Dunscore (1027) is a parish, with a village of the same name, in the western valley watered by the Cairn and Cluden, whose inhabitants are wholly engaged in agriculture. At Craigenputtock, on its uplands, Carlyle lived and wrote; at Ellisland Burns was a farmer; and at Lag Castle Sir Robert Grierson, the persecutor, dwelt. The Welshes of Dunscore and Craigenputtock are buried in the parish churchyard, Grierson of Lag in the old churchyard. Friars Carse, in the parish, was an early possession of the Monks of Melrose. It was the home of the Riddells of Glenriddell in the time of Burns, who there witnessed the famous contest for the whistle. (pp. 11, 16, 24, 42, 63, 123, 128, 140.)

Durisdeer (849) is a large inland parish in Nithsdale, occupied mostly by agriculturists and sheep-masters. A small kirk-town stands beside the parish church. In the churchyard Daniel Macmichael, the Covenanter, shot at Dalveen, is buried. Drumlanrig Castle is in the parish. The sites of Durisdeer and Enoch Castles remain prominent. Dalveen Pass, Enterkin Pass, and Kirkbride Church are favourite spots for visitors. (pp. 16, 97, 100, 106, 107, 109, 124, 140, 149.)

Eaglesfield (500) is a flourishing village in Middlebie parish, where many hands are employed in tailoring. Eaglesfield has a good library. Here William Lockhart the painter (1846–1900) was born. (p. 77.)

Ecclefechan (750) is a village in the parish of Hoddom (1258), once the seat of a gingham industry, and now the Mecca of devotees of Thomas Carlyle. Ecclefechan station (Caledonian Railway) is convenient for travellers to the Roman Camps of Birrens and Birrenswark in the vicinity. (pp. 20, 33, 128, 147.)

Enterkinfoot is a hamlet half-way between Thornhill and Sanquhar, in Durisdeer parish. It lies at the foot of the wild Enterkin Pass, the scene of a famous rescue of Covenanters from the dragoons of Claverhouse on 29th July 1684. The nearest railway station is Carronbridge (G. & S. W. Railway). (pp. 16, 125, 140.)

Carlyle's Birthplace, Ecclefechan

Eskdalemuir (392) is the parish with the largest acreage in the shire, namely 43,518½ acres, a district 13 miles long and nine miles broad. A well-defined Roman camp at Raeburnfoot (Overbie) and many circular forts remain. In 1908 an observatory completely equipped with magnetographs, seismographs, and other instruments for measuring atmospheric phenomena, was erected here because of the absolute serenity of the station. (pp. 3, 4, 10, 12, 21, 32, 42, 43, 57, 58, 60, 92, 97, 136.)

Gatelawbridge is a hamlet in Morton, where the quarry-men of an extensive sandstone quarry reside. (pp. 35, 76, 80, 145.)

Glencaple is a village in Caerlaverock parish, five miles below Dumfries, having a jetty where vessels of over 500 tons can moor and discharge, in 14 feet of water. It is a favourite resort for visitors. (p. 86.)

Gretna Green

Gretna or **Graitney** (1212) is a parish, with four villages —Old Gretna, Gretna Green or Springfield, Rigg of Gretna, and Brewhouses—and two railway stations. Being on the English boundary, it was famous for runaway marriages celebrated at Gretna Green. Gretna Green marriages are simple declarations of marriage made before witnesses, and sometimes recorded by the chief witness calling himself the celebrant. A marriage of this kind is still legal according to the law of Scotland, being styled an irregular marriage. Registration granted by order of the Court of Session or of the sheriff puts it into formal legal order for civil purposes; and the Marriage Notice (Scotland) Act only requires that one of the parties has his or her usual residence in Scotland, or has resided there for 21 days before the marriage. In "The Queen's Head" Inn at Springfield Lord Erskine was

Hoddom Church

married by David Lang in this old irregular way. The old "Gretna Hall" hostelry, now a private residence, "The Maxwell Arms," and the Tollhouse were scenes of marriage. These irregular marriages could be contracted in any house, the self-constituted "priest" being locally styled the blacksmith, because he welded the contracting parties. Battles were fought in Gretna parish. Here stands the famous treaty stone and landmark,

Clochmabenstane, and on the opposite shore at Burgh-by-Sands is erected a memorial to Edward I on the spot where he died. (pp. 20, 23, 35, 47, 50, 52, 56, 85, 97, 124, 128.)

Hoddom (1258) is a beautiful parish in Lower Annandale where St Mungo planted his see and Fechan had his church. The old castle of Hoddom is incorporated in a splendid mansion— the residence of Mr E. J. Brook. (pp. 14, 20, 29, 39, 40, 69, 84, 87, 88, 97, 100, 116, 117, 130, 147.)

Hollows, a small hamlet in Canonbie parish, takes its name from the fortalice of Johnie Armstrong, now repaired but not inhabitable, which stands above the Esk, four miles south of Langholm. The site of Gilnockie Tower is near by. (pp. 13, 116, 119.)

Holywood (865) is a parish, with a hamlet (69) of the same name, in Nithsdale. Some identify it with "the head of the wood in Caledon." Here stands the great circle of stones called "The Twelve Apostles." The old abbey, Dercongal, has disappeared together with a hospital built by Archibald Douglas, the Grim. Two ancient church bells hang in the belfry of the parish church. There is a quarry at Morrington. (pp. 45, 87, 97, 106, 138.)

Johnstone (751) is a parish in the Howe of Annandale, wherein is situated Raehills, the splendid residence of Mr Hope-Johnstone, proprietor of nearly all the parish. Lochwood Tower, now in ruins, the ancient family seat of the Johnstones, is built on a ridge overlooking a great morass on the west. The Lochwood Oaks are of great antiquity. (pp. 12, 20.)

Keir (538) is a parish, with a hamlet of the same name, beside the parish church, whose inhabitants are mostly engaged in agriculture, except those employed at Barjarg lime quarry. Here Kirkpatrick MacMillan, inventor of the bicycle, was born, lived, and was buried. Capenoch House and Barjarg Tower are in the parish. (pp. 16, 33, 100, 145, 150.)

Kirkconnel (2144) is a parish, with increasing village, in Upper Nithsdale, and includes part of the extensive coalfield, which at present employs 850 men and boys, and produces about 300,000 tons of coal annually. (pp. 28, 15, 30, 31, 32, 46, 76, 79, 100, 147, 149.)

Kirkmahoe (1080) is a well cultivated parish to the north of Dumfries. Its villages are Kirkton, Duncow, Sunnybrae, and

Langholm Parish Church

Dalswinton; its mansions, Duncow, Dalswinton, Newlands, and Isle.

Kirkpatrick-Fleming (1354) is a parish in the south-east district of the county, with a station on the Caledonian Railway. It lies within the red sandstone belt which was extensively quarried at Craigshaws, Branteth, Sarkshields, and New Cove. The last formerly employed 300 men and is still being worked. The chief mansions are Cove, Langshaw, Kirkpatrick, Mossknowe, Springkell, and Wyseby. (pp. 20, 104, 105, 115, 124, 143, 149.)

Kirkpatrick-Juxta (998) is a parish near Moffat. The villages are Beattock (with station) and Craigielands. (pp. 20, 138.)

Langholm (2930) is a burgh situated on the banks of the river Esk. The country around is charming. The chief buildings are the town hall, presented by the Duke of Buccleuch, the Hope hospital, built and endowed out of a bequest of £100,000 left by Mr Thomas Hope, a New York merchant, Langholm library and museum, academy, and Freemasons' hall. A statue to Sir Pulteney Malcolm stands beside the library, and an obelisk to commemorate Sir John Malcolm stands on Whita Hill. The chief industries are tweed weaving, which gives employment to 660 workers, tanning, till lately whisky distilling, and the sale of merchandise to an extensive district. There are six woollen mills, and large sales of sheep. (pp. 12, 13, 21, 22, 29, 31, 41, 60, 72, 73, 76, 77, 91, 105, 121, 128, 131, 133, 149, 150.)

Lochmaben (1056) is an ancient royal burgh in Annandale invested with much historic interest. Its lochs add distinction to the landscape, and the ruined palace of the Scottish kings, in the Castle Loch, is a vivid reminder of great national events. The original castle of the Brus stood on the Castle hill and is obliterated. The present castle, in parts, dates from before 1300. Tradition assigns the birthplace of King Robert I to Lochmaben. In the vicinity are four villages, called The Four Towns of Lochmaben, whose inhabitants have for many centuries been styled "The King's Kindly Tenants," being descendants of the vassals of King Robert the Bruce, who held their lands in copyhold, and being registered in the Rent-Roll of the King's representative, or castellan, at present, the Earl of Mansfield. These villages are

Greenhill, Heck, Hightae, and Smallholm. (pp. 4, 5, 6, 20, 24, 46, 47, 60, 66, 83, 92, 96, 106, 111, 112, 122, 124, 128, 130, 133, 135.)

Lockerbie (2455) is a small flourishing burgh town, beauti-fully situated, in the parish of Dryfesdale (3188), in Mid-Annan-dale. It is almost surrounded by the rivers Annan, Dryfe, and Milk, and offers a pleasant resort for anglers. Since the intro-duction of the railway in 1847, it has been converted out of a quiet rural village into a hive of industry. Its sheep and lamb sales are of great importance. Its modern buildings are hand-some and well-built; its streets broad and clean, and its water pure. It possesses a free library and higher grade school, besides golf links and other attractions of modern pleasure resorts. Fox-hounds and otter-hounds hunt here. In the vicinity is Dryfesands, the scene of a bloody conflict between Maxwells and Johnstones in 1593. Castlemilk, Jardine Hall, Elshieshields Tower are in the neighbourhood, together with the romantic ruins of Spedlins Tower. Lockerbie takes its name from the Locards, a family who anciently held land there—hence "Locarde-bi." The highest hill in the district, White Woollen or Quhyte Woollen, is pro-nounced White Ween. The Lockerbie "Tryst," or great lamb-fair, was formerly an annual event of much importance, and was held on Lockerbie Hill. Over 40,000 lambs were brought for sale. A local carnival was also held at the same time. (pp. 20, 60, 66, 68, 69, 72, 73, 76, 115, 120, 124, 128, 131, 133, 134.)

Middlebie (1722) is a rural parish in Annandale, with two villages, Eaglesfield and Waterbeck. Two old parishes, Penersax and Carruthers, are now united to Middlebie. In it are Birrens—a Roman camp, and Scotsbrig, where Carlyle's father farmed. The Carlyles anciently held lands here. (pp. 20, 66, 86, 97, 98.)

Moffat (2079), a police burgh, charmingly situated in Upper Annandale, 500 feet above sea-level, is approached by splendid

Moffat

roads, and has a station on the Caledonian Railway. For centuries Moffat has been famous for its mineral waters. The well affords waters, sulphurous and saline and having other medicinal properties, which are much in demand for curing gout, rheumatism, skin diseases and stomachic complaints. The Hartfell Spa is also a chalybeate well of similar character and popular for dyspeptic disorders. A large and magnificent Hydropathic, situated in

Renwick's Monument and Maxwelton Braes

lovely grounds on a commanding situation, adds distinction to the neighbourhood. The town with its broad, clean streets is a model for a health resort. A golf course, tennis courts, cricket grounds, bowling greens, pleasure grounds, and charming walks offer delightful rendezvous for tourists and visitors. The drives around are most attractive.

Off the main street of Moffat stands Moffat House, a seat of Mr J. J. Hope-Johnstone, Lord of the Manor. It is said to be a good specimen of Adam's style of domestic architecture.   (pp. 4, 12, 13, 18, 25, 29, 60, 72, 107, 109, 110, 124, 128, 133, 140.)

Moniaive (500) is a pleasant, quiet village in the upper part of the Cairn Valley in Glencairn parish (1410), nestling among the hills.   A light railway from Dumfries (G. & S. W. R.) has made this sequestered spot an accessible health resort, where hitherto agriculture and pastoral labour afforded a meagre employment to the villagers.   The owners of the land in Glencairn, as a general rule, reside in the fine mansions which adorn the district—Maxwelton, Crawfordton, Craigdarroch, Auchenchain, Townhead, and Caitloch.   James Renwick, the martyred Covenanter, was born at Knees, Moniaive, near the spot where an obelisk to his memory stands.   Several martyrs, shot in the parish, are buried in Glencairn churchyard.   There is a golf course at Moniaive.   (pp. 39, 103, 126, 128, 140.)

Mouswald (493) is a small agricultural and pastoral parish lying south-east of Dumfries.   Remains of castles, camps and cairns testify to its importance in ancient times.   St Peter was its patron saint.   Mouswald-Mains, or The Place, was the seat of the Carruthers.   Rockhall is the seat of the Griersons of Lag. (pp. 18, 29, 56, 107, 119.)

Penpont (831) is a parish and village two miles from Thornhill in Middle Nithsdale, whose inhabitants largely depend upon work created through agriculture.   (pp. 16, 30, 39, 100, 110, 135, 145.)

Powfoot is a favourite coast resort on the Solway, in Cummertrees parish, with golf course, bowling green, tennis courts, and other accessories of a summer watering place.   Angling is also procurable.

Ruthwell (770) is a parish, and village (100).   The Knights of St John of Jerusalem and Malta had a chapel, cemetery, and lands here. One of the finest standing crosses in Europe is preserved in the parish church. The village of Clarencefield (100) is in the parish. (pp. 29, 31, 35, 47, 50, 56, 57, 58, 60, 78, 88, 101, 102, 103, 114, 145.)

St Mungo, or **Abermelk** (573), is a parish south of Lockerbie which keeps alive the name of the great missionary, and patron saint of Glasgow, Munghu Kentigern (514–603), who

Town Hall, Sanquhar

fixed his see in Hoddom. The beautiful mansion of Castlemilk, the seat of Sir Robert Buchanan-Jardine, is in the parish. (pp. 63, 87, 133.)

Sanquhar (1508), a royal burgh, is situated in the valley of Upper Nithsdale. Its history goes back to an early period; and so early as 1296 we find "a new place," or stronghold, built at "Senewar," as it is then called. Sanquhar was a burgh of barony till 1598 when James VI erected it into a royal burgh. The present keep, sometimes called "Crichton Peel," was built in the fifteenth century by a Crichton. The market cross formerly stood on the street where an obelisk now commemorates two events which occurred there, namely the publishing of the Sanquhar Declarations. On 22nd June, 1680, Richard Cameron published the first declaration, which disowned allegiance to Charles II. On 29th May, 1685, James Renwick repudiated James VII and his government. There are small villages at Crawick Mill and Mennock Bridge. Agriculture, coalworks, brickworks, tileworks, and hosiery manufacturing give employment to the people. In summer Sanquhar is popular as a health resort, angling being easily obtained, and few restrictions existing to prevent visitors enjoying the hill scenery. Sanquhar has a golf course. (pp. 15, 29, 31, 32, 72, 77, 78, 79, 94, 98, 99, 107, 113, 114, 128, 130, 133, 135, 138, 140, 143.)

Thornhill (1169) is a lovely village in the parish of Morton (1820), in Middle Nithsdale, built on a high ridge, with broad streets, some of which form boulevards of lime trees. At the intersection of the streets stands a tall imposing market cross, under which fairs and markets were formerly held. It was a burgh of barony, and had a tolbooth, now a stable. Thornhill is the centre of a historic district. Three miles and a half north is the strong castle of Morton. Drumlanrig Castle, Closeburn Castle, Tibbers, and Enoch are also in the vale. The parochial buildings are of a very handsome character, and were gifts of the late Walter, Duke of Buccleuch. Morton Public School, one of the old parish schools, has produced many distinguished men, notably Joseph Thomson, whose monument stands outside the building. Angling facilities in the district are great. Regular series of drives to romantic places are attractive, and now Thornhill is a health resort. In the vicinity of Thornhill are the great sandstone quarries of Gatelawbridge, Newton, and Closeburn. (pp. 12,

16, 33, 35, 41, 43, 73, 76, 77, 100, 103, 104, 105, 107, 113, 124, 125, 147.)

Tinwald and **Trailflat** (728) is an agricultural parish, five miles north-east of Dumfries, in which stands Amisfield Tower, the ancient seat of the Charteris family, of whom Sir Thomas was Chancellor of Scotland in the thirteenth century. (pp. 18, 88, 93, 145, 147.)

Morton School and Schoolhouse, Thornhill

Torthorwald (788) is a parish and village, west of Dumfries, whose ruined tower was formerly the seat of Kirkpatricks, and afterwards of Carlyles. (pp. 18, 107, 112.)

Tundergarth (399) is a parish east of Lockerbie, in which is a circle of stones known as "The Seven Brethren." Traces of the paved Roman road are seen here. (p. 20.)

Tynron (309) is a small but romantic hilly parish, watered by the Shinnel, and dominated by the commanding triple-ditched fort, called Tynron Dun, to be seen from every point in Mid-Nithsdale. The Roman road is said to run from the Dun to Drumloff, crossing the Shinnel above Stenhouse. In this parish Robert the Bruce hid. In the churchyard lies the body of William Smith, a Covenanter, shot on Moniaive Moss in 1685. James Hogg, the Ettrick Shepherd, tended sheep in this parish. (pp. 11, 16, 30, 96, 109.)

Wanlockhead
(*The highest houses in Scotland*)

Wamphray (369) is a parish, stream, and station (Caledonian Railway) in Annandale. The Roman road is traceable through it. Wamphray Glen is noted for its beauty. In the old tower of Wamphray lived "The Galliard," a Johnstone, of whose fighting retinue it was said "the Wamphray lads are kings of men." A famous minister of Wamphray in the seventeenth century, and a voluminous writer, was John Brown. Exiled to Holland, he there harassed the government of Charles II by his pungent pen. (pp. 12, 20.)

Wanlockhead (624) is a *quoad sacra* parish and village on the confines of the county, formerly in Sanquhar parish. It is famous for its lead-mines, employing 250 workmen. A light railway connects this village among the hills with Leadhills and Elvanfoot, in Lanarkshire. The station stands 1384 feet above sea-level. In the churchyard is interred Professor William Hastie, a remarkable scholar. (pp. 8, 29, 30, 76, 79, 129, 138.)

Westerkirk (393) is a pastoral parish to the north-west of Langholm. An antimony mine at Glendinning was worked up to about twenty years ago. Westerhall, a well-wooded estate, an old seat of the Johnstones, with salmon fishings worth £300 a year, was sold in portions in 1911. (pp. 21, 80, 143.)

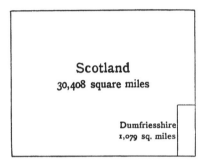

Fig. 1. Area of Dumfriesshire compared with
that of Scotland

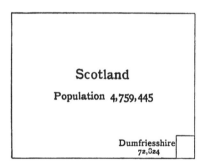

Fig. 2. The population of Dumfriesshire compared
with that of Scotland in 1911

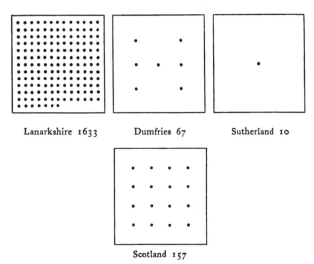

Lanarkshire 1633   Dumfries 67   Sutherland 10

Scotland 157

Fig. 3. Comparative density of Population to the
square mile (1911)

(*Each dot represents ten persons*)

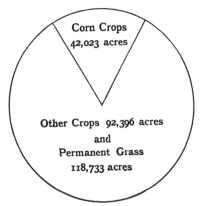

Fig. 4. Proportionate area under Corn Crops in
Dumfriesshire in 1910

DIAGRAMS 175

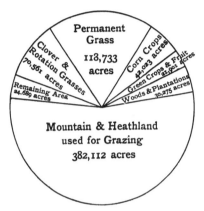

Fig. 5.   Comparative areas of land in Dumfriesshire
in 1910

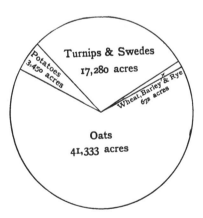

Fig. 6.   Proportionate areas of land producing Corn,
Turnips, etc., and Potatoes in 1910

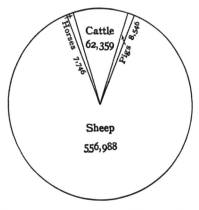

Fig. 7. Proportionate numbers of Sheep, Cattle, Horses
and Pigs in Dumfriesshire in 1910